Simply Learning, Simply Best!

Simply Learning, Simply Best!

INTERNATIONAL BUSINESS ENGLISH

外貿業務英文

劉美慧 ◎著

從入門到精通，一本搞定外貿全流程

用專業英文和全世界談生意

【修訂三版】

完整的外貿業務操作流程，

高效提升外貿業務職場競爭優勢

進出口貿易從業人員、

各大國貿系所學生、教學輔助教材，都適用！

Author 作者序

在我的年代還有聯考這件當年許多人嫌棄、現今大家懷念的制度，當我還是一整個秀外慧中的青春無敵高中女學生時，夢想就是要進大學唸外文系，但繳交志願卡的前一天晚上跟同學講著電話，這位同學跟我說一定要唸商才有前途……就這樣，我臨時改換了志願順序，最後進了政大財稅系，唸了之後才知道它隸屬於法學院……

大學畢業後，進了大東亞藥品貿易公司，當了英文業務祕書，這工作無關財稅專業，我所圖的是它工作會用得到英文！後來又到了台儀集團，做了總經理秘書……因為心中一直有想唸研究所的想法，所以辭了職，結了婚，考進了師大國際人力教育與發展研究所，而選定它報考的主要原因在於它的課程設計為全英語授課。

2008年時，北藝大的周英戀老師找了我當「國際技能競賽一花藝職類」的隨行翻譯，為了想讓自己將這份工作盡力做好，所以決定報考師大翻譯所，考上了，畢業了，也當了四屆的國際競賽隨行翻譯……

說了這些，是要告訴你生命的無常……哈！不是的啦，我要說的是，我們生命中的軌跡不見得是我們原先所設想的，也不一定是我們可以想見得了的，但我們鐵定可以做到的是：做好當下的每一件事！

就舉外貿英文的工作與學習來說，若是我們能盡力將每一天工作中所接觸到的外貿事務做得完整，將收到的每一份國外Email與文件看個徹底，有好奇心，有求知慾，那麼，用這樣的態度處事，每完成了一份工作，我們就可以成為那份工作的專業人士！而當我們做的事愈來愈多，知識與經驗愈來愈通透純熟時，若有新的工作交派給我們，我們自能有信心地接下挑戰，也可以有能力且有辦法地去完成任何可能與不可能的任務！共勉之！

劉美慧（Amy）

Foreword 前言

這本書要說的主軸就是「外貿」，要從事對外出口貿易的工作，要熟悉國際貿易的實務操作，一定要具備外貿的專業知識，也一定要對英文這語言有一定的駕馭能力。所以，在這本書裡，我將帶著你先來練練業務作業上的基本功，從報價、下單訂購到出貨這些每天不斷重複發生的外貿工作來談，接著會再領著你修煉進階的內力功夫，看看行銷活動與業績檢討上的溝通情境與關鍵要素。

不過，外貿業務、行銷與業績的層面那麼廣，我們要從哪兒切入？要著重哪些部分呢？舉個例來說，我們會說到價格的結構，但不會細細地從企業內部的生產要素與成本管控來說，而是會說到成本、經銷商價格、零售價格，以及加成與利潤率這些要素的關聯，為什麼呢？因為企業內部的生產或是成本管控，你會是用中文跟主管與同事說，而經銷商價格與零售價格都是你會跟國外經銷商與客戶用英文談到的事項……所以囉！這本書的骨骼架構是對外貿易業務的作業要素，而英文就是它的肌肉組織！

在這本書裡所列的「對話」情境，就是你會跟外國友人談到的外貿業務內容，書裡列的「說三道四，換句話試試」，就是要讓你看看跟外國友人如何用不同的話談同類型的事，書裡的「關鍵字急救站」，要帶你看看這些基本關鍵字詞的內涵，很多時候有的英文字我們以為看過了、看熟了，就是懂了，但其實不然，因為我們還沒知其所以然，而從這些關鍵字詞一路探究下去，讓你懂得的也不僅僅只是英文，更有相佐的貿易專業知識呢！此外，書裡還有一個重頭戲小單元，那就是「術語直達車，專業補給站」，光瞧上「術語」與「專業」這兩個詞兒就知道這個小單元的份量了吧？！不過，請別擔心，我們可以「深入」談，但也懂著如何「淺出」，絕對讓你順順地讀

下去，知識就這麼自然地吸收進去了呢！書裡每一節的最後一個小單元，就是要讓你動手練習一下的「Show Time！換你上場！」當我們看得順，都順到邊看邊點頭時，其實還真的只是看順眼，了然於心，還沒到真正入了心呢！所以，到了這個練習的小單元，就請真的拿支筆，動手寫寫吧！

那這本書說的是對外貿易，所以就是適合工作上會處理到「出口」業務與行銷的主管、業務、秘書與業務助理嗎？是！但不只是這些人而已喔！若你從事的是與「進口」有關的工作，這本書也很適合你來讀，為什麼呢？出口與進口是國際貿易行為的兩個方向，這本書從出口廠商與國外溝通的方向來寫，說的事項與要點，同樣地也就是進口廠商與國外溝通的內容！所以，只要你的工作與國際貿易有關，這本書都會是一本能夠帶著你提升能力的書。

這本書寫的是外貿知識，寫的是英文，但寫的其實也是看事情的角度，遇事處理的態度。若是我們能處處留心，事事上心，那我們就算是從日常工作處理上的小事情中，也能看出它之所以會發生的原因，也能想出它的邏輯所在，也可窺得它與其他事件的牽連關係，而當我們對事理瞭解的愈來愈清楚，在處理上就能愈來愈快，愈做愈好！外貿業務的許多知識其實就隱藏在你每天的工作中，若你接觸到了但不多想，知識就不會顯現出來，若你想到了但覺得無所謂而把它就此忽略掉，那麼你就無法累積你的知識厚度。對外貿知識來說是如此，就英文能力來看也是如此，在你看這本書時，若你能仔細看過每一個英文字詞，跟著書裡說明字詞的邏輯走，讓它在你心中多繞個一圈，在你嘴上多唸個一次，讓你對這樣的功夫天天天天練習個幾回，若是你肯這麼做，那麼外貿英文也就不會難到哪兒去了呢！

音檔下載 

Contents 目次

Contents 目次

Part 1
業務基本功

Chapter 1 篇章簡述

在國際貿易的銷售規劃上，知道如何定價、用何種方式出貨，以及要求客戶以何種方式付款，可都是原廠或供應商考量的要點，而其中居上位的超大要點，就非價格莫屬了！

價格這事可是茲事體大，不管你銷售量多高、市佔率多漂亮，若是價格訂得太低，可能忙再多都是做白工，能夠賺到的可能只有經驗，而沒有利潤！

在訂定價格時，原廠或供應商內部的成本與策略目標，以及外部的市場競爭與市場需求，都是得要考量的因素。原廠或供應商訂定出口價格時，可從國內定價為起始點，由此基準往上加入出口貿易的成本，再算上利潤，即可得出產品的定價。

有了定價之後，還得要好好算一算相關的折扣率。要拓展出口貿易業務，第一步通常會是尋找國外當地的經銷商，因此，在確立定價後，經銷商折扣也要定出來，接著也得盤算一下數量折扣，看是可用什麼樣的折扣設計來吸引買方在量上面再衝刺一下，多買一些。

而在原廠或供應商訂出定價與經銷商價格之後，實際上還有好些個行政作業的費用得要理一理，包括包裝費、處理費、運費、保險、出口稅捐、文件費、銀行手續費等等，這些都要先確定下來，之後才能回答經銷商與客戶提問的各種與價格有關的問題。

經銷商和客戶提出詢價時，可是各式各樣的問題與交易條件要求都會傾巢而出，例如「請報給我們FOB價格」、「出貨若是由你們安排，運費會是多少」、「價格能否更優惠一些」……這些提問都跟價格有關，也都會影響原廠或供應商的報價。像是關於FOB價格這事兒，FOB屬於交貨條件的一種，而在Incoterms／國貿條規中，一共列出了11種交貨條件，界定並說明了國際貿易中買賣雙方應負的責任、成本與風險。在客戶不同的交貨條件要求下，原廠或供應商就得理出各種相關的費用，以調整報價。

當經銷商和客戶發來詢價時，經常會有特價或專案價格的要求，而原廠或供應商對於能有什麼降價空間、最低價格可下到什麼樣的水準，可也要花些功夫評估一下，而在評估的同時，也有需要瞭解一下經銷商的報價與價格結構。

在經銷商和客戶下單前，會要求原廠或供應商提供的文件包括了與價格有關的文件，如報價單與Proforma Invoice，也包括了與產品有關的文件，例如Specification／產品規格書與Operating manual／操作手冊等。產品資料依產業、產品不同而異，而報價單與Proforma Invoice可就是跨越各產業都能通，都有其必要要素，在表格上都可見到這些要素的一席之地。

在這一章裡，我們會從報價種類、折扣種類、交貨條件、運費、經銷商報價結構一路看下去，最後再來實地看看報價的表格，讓你把價格這事兒徹底瞭解個夠！

搞定報價

1-1 報價種類

Janice: Hi, I'd like to speak to Mike, please.

Mike: This is Mike. Is this Janice?

Janice: It sure is! We're currently considering placing an order for your Super 101 Collagen Chewable.

Mike: Oh, it's our newly released product. Its price is 60 USD for 120 tablets per bottle.

Janice: It should be the list price, right?

Mike: Yes, its transfer price is 48 USD.

Janice: What do you mean by "transfer price" ? Is it the one after a 20% distributor discount that you offered us?

Mike: Exactly! You could have our standard distributor discount for all products except the 500 series. None of the 500 series items are eligible for any discounts.

Janice: Got it. Is Super 101 Collagen Chewable available?

Mike: Yes, we have plenty in stock. If I could receive your order by Thursday, I can get it out the door to you this Friday.

Janice: Perfect! We'll decide the quantity and then put in an order for this product!

Mike: Great! I'm looking forward to it!

背景說明	
人物	Janice：經銷商業務人員 Mike：原廠業務人員
主題	確認價格，準備下單訂購： 定價 vs. 移轉價格／經銷商價格

譯文

珍妮絲：嗨，麻煩請找麥可。

麥　可：我是麥可，妳是珍妮絲嗎？

珍妮絲：沒錯！我們打算要下單訂購你們的「超級101膠原蛋白嚼錠」。

麥　可：喔！這是我們最新推出的產品，每瓶120顆的價格是美金60元。

珍妮絲：你說的是定價，是吧？

麥　可：是的，它的移轉價格是美金48元。

珍妮絲：這個「移轉價格」是什麼呢？是加計了給我們的20%經銷商折扣後的價格嗎？

麥　可：完全正確！除了500系列之外的所有產品，你們都可享有標準的經銷商折扣。500系列的品項是不適用任何折扣的。

珍妮絲：瞭解。那「超級101膠原蛋白嚼錠」有現貨嗎？

麥　可：有的，我們還有不少現貨。如果我這星期四之前能收到你們的訂單，那我就可以安排星期五出貨。

珍妮絲：太棒了！我們會將訂購數量決定出來，然後就下訂單給你囉！

麥　可：太好了！那我就等著囉！

說三道四，換句話試試

例一 | 報上定價，說明經銷商折扣數、折扣價，再另行送上價格表。

例句 Our list price for AL61 Microscope is 800 USD and you could receive our 30% distributor discount for a price of 560 USD per kit. Attached please also find our distributor price list. If you have any questions, please feel free to contact me.

譯文 AL61顯微鏡的定價為美金800元，您可享有30%的經銷商折扣，所以折扣後的單價為美金560元。在此也附上我們的經銷商價格表，若您有任何的問題，還請直接與我連絡。

例二 | 逐條列出報價編號、報價效期、定價及經銷商折扣價。

例句 Please find below the price for the requested item:

Quote Number: CDC0501 (This quote is valid within 30 days.)

List Price: 800 USD/Kit

Transfer Price: 560 USD/Kit

If you have any questions, please contact us at any time.

譯文 您所要求之品項的價格如下所示：

報價編號：CDC0501（此報價30天內有效）

定價：美金800元／組

移轉價格：美金560元／組

若您有任何的問題，請隨時與我們連絡。

例三 | 定價、折扣、移轉價格，一起列表說明，外加報價索引號碼。

例句 You could have your typical distributor discount for the inquired product. The table below reflects the price with the discount applied in the Transfer Price column. Your reference quote number will be Q1010.

Cat. #	Description	Size	Qty.	US List Price	Discount	Transfer Price
3008	AL61 Microscope	kit	1	$ 800	30%	$ 560.00
					TOTAL	$ 560.00

譯文 您們所詢問的產品可適用標準的經銷商折扣，下表中「移轉價格」一欄就是反映了折扣後的價格。報價索引號碼為Q1010。

型號	品名	尺寸	數量	美金定價	折扣	移轉價格
3008	AL61 顯微鏡	組	1	$ 800	30%	$ 560.00
					總計	$ 560.00

例四 | 列出細部組件的定價，網路下單結帳時會自動適用經銷商折扣。

例句 Below is a list of the components for this kit with list price. Your distributor discount will be applied at the checkout.

譯文 如下清單列出的就是這一組產品所含組件的定價，結帳時會加計您的經銷商折扣。

關鍵字急救站

在提出詢價與回覆報價時，「quote」這個字可就是我們常常會見面的好朋友了。先來說說它的詞性，「quote」可當動詞，也可以當名詞，當動詞時，意思為「to tell someone what price you would charge them to do a particular piece of work」，也就是對你要提供的產品或服務估個數，報個價，像是說著：「Please quote your best price for this instrument.」（對於這台儀器，請報給我們您最好的價格。）「quote」也可當動詞，而動詞quote加了-tion的名詞字尾，變成quotation，也是個名詞，所以囉，當你要給對方報價、附上報價單，或是告訴對方報價索引號碼時，所用的都是名詞quote或quotation，有時還會quote跟quotation左右開弓，在幾句話當中，一下子用名詞quote，一下子又用個quotation，這…這…這不是自找麻煩嗎？喔不！這是生活情趣！啊～也不是啦！這其實就是寫作用字上的變化而已，而有時也不是刻意想變化，只是e-mail中順順地說著quotation，但在說到報價單單號時，因為報價單的系統本來就將單號設為「Quote No.」，所以還滿常看到這裡quote、那裡quotation的狀況。

例句一　To request a quote for a particular product or service, please follow the links or complete and submit the form below.

譯文一　若您需要某項產品或服務的報價，請依連結所示，或是填寫並提交如下表格。

例句二　Attached is our quotation. Please refer to the Quote No. when placing your order on this customer's behalf.

譯文二　在此附上我們的報價，當您替客戶下單時，請加註此報價單單號。

術語直達車，專業補給站

經銷商價格 vs. 移轉價格

原廠報價時，常會跟經銷商說到「The transfer price for your inquired product is...」，我們看多了、寫熟了，也說順了，知道我們對經銷商說的這個transfer price／移轉價格，也就是distributor price／經銷商價格，那為啥要用transfer呢？一說了「移轉」，怎麼聽起來好像就沒那麼多銷售、謀利的氣息了呢？是的，你的直覺沒錯，但有了直覺之後，請就繼續挖掘，因為我們要用來培植實力的知識，通常就隱在我們所接觸的事物當中，沒多想就沒得領悟，有探索才會有所得！而知識就是這樣累積而來的啊！

好了，我們先來看看transfer的意思：「the process of moving, or of moving someone, from one job, department, or office to another in the same organization」，而transfer price此價格的定義是「the price charged when one segment of a company provides goods or services to another segment of the company」（公司的一個部門提供產品或服務給另一個部門所收取的價格），看到關鍵點了嗎？就是「same organization」、「same company」，企業提供產品或服務給自家的部門或關係公司時，所銷售、購買的價格與公司自身的利潤、成本計算有關，也會與應繳稅額有關，所以呢，公司也就可以從這一塊來運作，以減低稅金，達到節稅的效果。

咦？可是原廠跟經銷商又不同屬一個企業或集團啊？確實沒錯，不過原廠與經銷商在合作關係上的確也可說是生命共同體，原廠依靠經銷商在當地市場運籌帷幄，拓展業務，提高品牌曝光度，拿下市佔率，而經銷商依賴原廠的貨源提供，轉手銷售，提供服務，以獲取利潤。此外，就價格水準來說，原廠給經銷商的價格會在有利潤的條件下，低於零售價格，而經銷商以較低價取得產品或服務之後，也還能在有利潤的基礎上，再銷售給其客戶，而這也就是原本所定義之移轉價格的定價策略與操作了。

那麼，當你要報價給經銷商時，可不可以直接說報的是distributor price呢？當然可以！那說是transfer price呢？當然也對！就看你那一天的心情想寫distributor還是transfer囉！

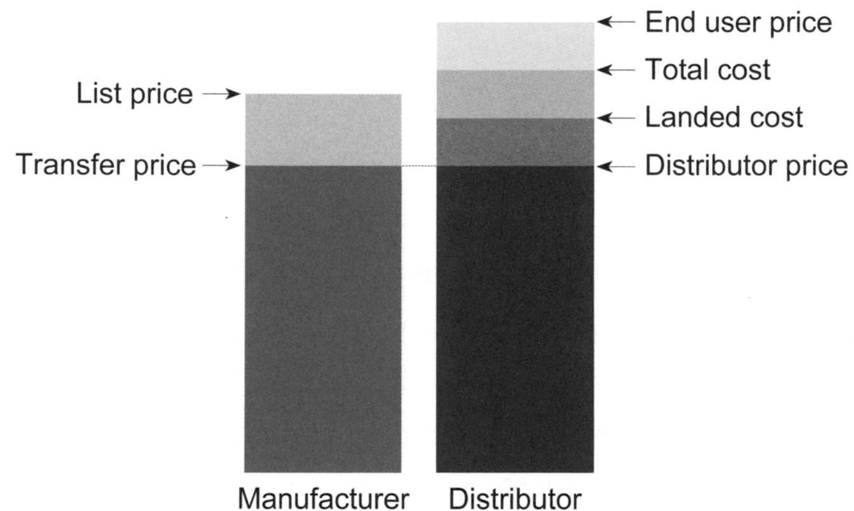

註：Landed cost／落地成本 — 1-2 有超詳盡的說明，歡迎先翻過去瞧，或是等著瞧囉！

最後，讓我們來說一下transfer的發音，它就是那一類當動詞與當名詞時的重音在不同音節的字。當名詞時，它的音標是 [`træns fɝ]，重音在第一音節，而這裡說到的transfer price，取的即是其名詞用法，所以唸的時間重音要落在第一音節。當它是動詞時，音標為 [træns`fɝ]，重音就跑到了第二音節。

在國貿實務上，transfer還真的算是常見字，除了跟經銷商報價時會用到transfer price之外，在談到付款、匯款時，像是原廠要求預匯貨款，不接受信用卡付款，就會這麼說：「We request an advance payment by bank wire transfer (not credit card payment)」【我們要求預先匯款（不接受信用卡付款）】，而在報價單、Invoice上的付款條件處，更是會看到「Wire Transfer」這個付款方式，下方就會帶出收款銀行的帳戶明細。這裡說到的兩處transfer用法，所用皆是名詞，要是寫出的是這樣的句子：「I have attached the information you need to transfer the payment for AL61 Microscope.」（我附上了您要匯AL61顯微鏡貨款所需要的資訊），這裡所取的就是transfer當「轉帳、匯款」的動詞用法囉！

Show Time! 換你上場！

❶ 這個產品的定價為美金800元，您可享30%的經銷商折扣，所以折扣後的單價為美金560元。

Our ____________ price for this product is 800 USD and you could have our 30% ____________ discount for a price of 560 USD per kit.

❷ 除了500系列之外的所有產品，你們都可享有標準的經銷商折扣。

You could ____________ our standard distributor discount for all products ____________ 500 series.

❸ 在此附上我們的報價，當您為此客戶下單時，請加註此報價單單號。

____________ is our quotation. Please ____________ ____________ the Quote No. when placing your order on this customer's behalf.

❹ 我們所報的價格在市場上仍然具有競爭性，因此，我們不認為我們有必要降價

The price we ________ still remains ________ in the market and therefore we don't think it's necessary for us to ________ our price.

來對對答案

❶ list；distributor

❷ receive / have / get / apply；except

❸ Attached；refer to

❹ quote；competitive；lower

說說文，解解字

❶ 定價就是list price，list是列表或目錄，所以那些會標在表上、型錄上的價格，就是定價。list price跟distributor price都是「名詞＋名詞」的複合名詞，第一個名詞當修飾語，具形容詞功能，通常以單數表示，所以不會有lists price或distributors price這樣的詞組出現喔！

❷ 要表示「取得」折扣的說法有好幾個，receive / have / get / apply a discount都可說！當你在寫Email時，請習慣要求自己換個習慣，對自己常用的字，也來換換說法，像是替換個詞兒，或是主動換被動都是很好的練習。例如在這一題中，原先寫的是以You為主詞的主動語態，那我們就可將discount換為主詞，改為「Our standard distributor discount could be offered to you for all products except 500 series.」。

❸ 要在Email中發個附件，說法可有好幾種，都是在attach這個字上變來變去，詞序換來換去。題目中的「Attached is our quotation.」，為從原句序「Our quotation is attached.」將attached這個補語移到句首倒裝而成，而除了題目的說法之外，還可以這麼說：

➡ Attached please find our quotation.

➡ Please find the attached quotation.

➡ Our quotation is in the attachment.

➡ Please find our quotation in the attachment.

❹ 此題的remain為連綴動詞，意思為保持、仍是，是在remain後頭接competitive這個形容詞。同樣的用法還有 remain unchanged／維持不變、remain silent／保持沉默、remain confident／仍然有信心。

搞定報價

1-2 折扣種類

Hank: Good morning. I'm calling for Ann Chen.

Ann: It's Ann speaking. What can I do for you?

Hank: Ann, this is Hank Jones. I got your quote for our inquired 8 kits of LMA Enzyme yesterday. Is 400 USD our distributor price?

Ann: Yes, I included 20% distributor discount in this quote. Is there any problem?

Hank: Our customer said our price this time is higher than competitor's! Could you lower the price?

Ann: In that case, we encourage your customer to make his purchase in larger quantity. If you order 10 kits or more in a single order, you will receive 25% discount off the list price!

Hank: That sounds attractive! I will try to convince my customer to order bigger volume of kits so as to get this quantity discount.

Ann: Great! By the way, we also have 3 short dated kits which will be expired on 7/10/2023. If you could accept this expiry, I could give you a special price of 250 USD.

Hank: Thanks for the information. I'll discuss with our customer and get back to you on this. Hope to give you a larger order!

Ann: Thanks, Hank. You're really our great partner!

背景說明	
人物	Hank：經銷商業務經理 Ann：原廠客服經理
主題	折扣大一點： 經銷商折扣、數量折扣、短效期產品特價促銷

漢克：早安，請找陳安。

安：我就是，請問有什麼事嗎？

漢克：安，我是漢克．瓊斯，關於我們所詢的8組LMA酵素，我昨天有收到妳的報價了，請問妳報的美金400元是給我們的經銷商價格嗎？

安：沒錯，這個報價有包含了20%的經銷商折扣，怎麼了嗎？

漢克：客戶說我們這次的報價比競爭者高耶！你能不能降價呢？

安：這樣的話，我們建議你的客戶考慮一下增加訂購的數量，如果你一次下單訂個10組或更多，我們可提供定價25%的折扣。

漢克：聽起來還滿吸引人的！我會去說服客戶多訂一些以取得折扣。

安：太好了！對了，我們另外還有3組短效期的現貨，效期是到7/10/2023，如果你可以接受，我可給你美金250元的價格。

漢克：謝謝妳告訴我這個資訊，我會跟我的客戶討論後再回妳，希望可以給你大一點的訂單！

安：謝謝你了，漢克，你真是我們的好夥伴！

例一｜有折有折！一波三折！

例句 Our typical list prices for these kits are: KT-721 490 USD/kit; KT-722 640 USD/kit; KT-723 490 USD/kit. For these three kits you would receive a quantity discount as follows:

10 kits: 10% off　25 kits: 15% off　50 kits: 20% off

So if you purchase 9 kits of each item you will receive a 15% discount off the list prices for the items, because you ordered more than 25 kits in a single order.

譯文 對這幾組產品，我們一般所給的定價為：KT-721 美金490元 /1組; KT-722 美金640元 /1組; KT-723 美金490元 /1組，這三組產品可給您如下的數量折扣：

10組：9折　25組：85折　50組：8折

因此，若是你每個品項各買9組的話，你就可享有定價的85折，因為你一次下單所訂購的數量超過了25組。

例二｜新推出數量折扣方案，多買多折！

例句 We recently instituted a volume discount program as follows to make it easier for you to get even greater discount values on your purchases. If you are planning your projects for this year, consolidate your orders and enjoy greater savings than ever before!

➡ 10 kits or 2,500 USD: 10% off　20 kits or 5,000 USD: 20% off

譯文 我們最近訂定了如下所列的數量折扣方案，讓您採購時更容易享受到更大的折扣。若是您正在規劃今年的專案，請一次訂多一些，讓您比以往省更多！

➡ 10組或 美金2,500元 = 9折　20組或 美金5,000元 = 8折

例三 | 價格已探底！沒法折！

例句 The distributor price of 247 USD is the lowest price that we are able to go for any quantity of kits. Our distributor pricing is our absolute lowest price. Quantity discount pricing is taken off of our list price. Since the list price for the AML0712 is 407 USD, 10% quantity discount pricing would be 366.93 USD. So you can see the distributor price of 247 USD is a much lower price. Please let me know if you have any more questions.

譯文 經銷商折扣價美金247元已是我們對任何數量所能給的最低價了，經銷商價格絕對是探底價，因數量折扣是自定價算下來，像是AML0712的定價是美金407元，加計10%數量折扣後的價格為美金366.93元，所以美金247元的經銷商價格其實比這個價格低很多。如果您還有其他的問題，再請告訴我們。

例四：此處不折，他處折。

例句 I'm sorry to tell you that there is not a discount on the OLC line of products. They are marked as low as we can price them. However, you may find some tier pricing on some of the other products, as shown on our website, which are priced according to your ordered quantity.

譯文 抱歉，OLC產品線的產品並沒有折扣可供，它們的價格已是我們所能給的最低價，不過，您可以在我們網站上看到我們有些其他的產品有分層訂定的價格，那就是根據您訂購的不同數量所定出的價格。

關鍵字急救站—Discount

買家若是行家，一定會想方設法地讓賣家釋出一些「discount」，讓自己買得滿意。而賣家若是行家，也一定會挖空心思，規劃一些discount program／折扣方案，吸引買家下單或訂多些。「discount」這個字可當名詞，也可當動詞，詞性不同，發音的重音音節也就不同，名詞時唸為[ˋdɪskaʊnt]，重音在前，動詞時唸為[dɪsˋkaʊnt]，重音在後。

我們先來說說名詞。客戶常常會問「Is there any discount?」，原廠常會回答的是「We can give you a discount of xx% if you purchase xx or more kits.」，接下來原廠就會告訴客戶他們的quantity discount program／數量折扣方案，或是discount structure／折扣結構。原廠要給客戶特價時，一般也會給客戶一個discount code／折扣編碼，說著「Discount code should be referenced when placing your order to receive the discount.」，請客戶下單時加註，這樣原廠處理訂單時才會以特價來算。原廠若要吸引客戶多訂些貨，discount可是個利器，像是「Ordering both products in one order will have a 10% discount on your purchase.」，說著兩種產品一次訂購，則可享10%的折扣，或者像是「Further discounts are available if you reach certain sales targets.」，吸引客戶達成一定的銷售目標，到時候就可加碼放送折扣呢！「discount」常見的動詞搭配詞語有discounted rate與discounted price, 像是原廠會說「We are offering a selection of products at a discounted rate.」，表示特選了一些產品，以折扣價提供給客戶，或說「Here is the chance to purchase our short expiration inventory at a discounted price.」，告訴客戶以折扣價購買短效期存貨的好機會來了，這些都是要運用折扣來集結人氣、吸引買氣！

術語直達車，專業補給站

在國際貿易上，我們會看到好多種不同的discount説法，到底有哪些呢？就讓我一併列出來讓你掉個下巴、驚訝一下吧：

- Commercial Discount／商業折扣
- Trade Discount／貿易折扣
- Distributor Discount／經銷商折扣
- Quantity Discount／數量折扣
- Volume Discount／數量折扣
- Cash Discount／現金折扣
- Settlement Discount／結帳折扣

這麼多折扣啊！「那……那……我還看過Special Discount！」，對對對，只要你説得出來都算對呢！在你看到上面出現兩個中文的「數量折扣」時，你應該有感覺到有些折扣的內涵是一樣的，只不過是名稱有異吧！是的！我們現在就來用心分類一下，好讓你開始對折扣有點感覺，也可感受一下客戶對折扣的感想！當客戶多要了幾次折扣後，你就會對它有依賴的情感。我們人啊，一有了情感之後，就容易情緒化，所以之後得到它時，會感激！得不到它時，就會感慨啊～ 哈！好了好了，在我們練習完「感」字的造詞之後，就可以乖乖來看看折扣「discount」的造詞囉！

折扣是什麼呢？就是賣方在產品原定的價款上，給予買方價格的優惠。在商言商，商場上、國際貿易中所説到的折扣，都跟商業有關，所以都可統稱為Commercial Discount／商業折扣。商業折扣通常可分為三大類：

一、Trade Discount／貿易折扣
Distributor Discount／經銷商折扣

將一個產品自定價、型錄價格或零售價格扣除部分金額後，銷售給中間商，以換取中間商銷售其產品給最終使用者所提供的服務，像是倉儲、出貨處理等。此處所扣除的金額即是貿易折扣，因折扣給的對象是中間商，即為經銷商，因此也就是經銷商折扣。

二、Quantity Discount／Volume Discount／數量折扣

原廠提供數量折扣的目的在於吸引買方大量訂購，讓買方以量大來換取較低的價格。經銷商除了經銷商折扣之外，也會以大訂購量來要求原廠提供額外的數量折扣與特價。在原廠網站中常會推出、提供數量折扣，若是針對的對象是最終使用者，經銷商並不適用，就會這麼說：「This promotion is only available to end users. Distributors are not included.」（此促銷方案僅適用於最終使用者，經銷商不在適用範圍之內）。實際上，這些提供給最終使用者的折扣，大多還是比經銷商折扣來得小，因此經銷商當然也不會去使用到這樣的數量折扣。

三、Cash Discount／現金折扣
Settlement Discount／提前付款折扣、結算折扣

這類的折扣說了現金、又講了結帳，都跟錢有關，所以，想當然耳，它們就是為了要吸引買方快快付款而給定的折扣，像是這樣的條件：「2% off the bill if payment is made within 10 days」（若在十天內付款，則可享帳單金額2%的折扣）。

除了上述三類折扣之外，就沒有別種折扣了嗎？當然還有囉！像是要說所提供的是僅此一次，下不為例的特價，就可能會說到「You have been approved for a 50% one-time discount.」（我們已核准提供給您50%的一次性折扣），這樣的折扣所對照的就是「typical distributor discount」這種可以天天有、天天給的典型經銷商折扣了。另外還有所謂的「Seasonal Discount」／ 季節性特價，這種特價不是在產品正值旺季時所給，而是要鼓勵買方在淡季時好好撿一下便宜，提早採購、預作準備，這樣也可以讓原廠減輕倉儲的壓力，也能事先安排生產等事宜。

沒折沒扣沒留心，有折有扣有吸睛！折扣這事是談成訂單前的一件大事，客戶想要、經銷商想求，而原廠想要給出一個減價最少但又可達到效果的特價，之後就可歡喜等候訂單。所以，折扣請好好談，求的人請將資訊給齊，給的人請仔細評估後給出一個具競爭性的價格或是充足的理由。

此外，在你看著與折扣相關的Email內文時，請睜大眼睛看看所給的折扣率是off the list price／從定價算下來，還是從經銷商折扣算下來的off the distributor price或additional discount／額外折扣？這點數字小事，可得要仔細算計算計，若是閃神看錯了，之後可就不是急著跳腳跳幾下就沒事的呢！

Show Time! 換你上場！

❶ 如果您一次下單訂個10組或更多的數量，您就可取得從定價算下來25%的折扣。

If you order 10 kits or more in a single order, you will receive 25% discount ________ ________ ________ ________.

❷ 我們最近制定了如下所列的數量折扣方案。

We recently instituted a ________ ________ ________ as follows.

❸ 我們的經銷商價格絕對是探底價，而數量折扣是從定價算下來的。

Our distributor pricing is our ________ ________ price, while quantity discount pricing is taken off our list price.

❹ 請注意，若是您達到一定的銷售目標，則還可提供給您更多的折扣。

Please note that ________ discounts are available if you reach certain ________ ________.

來對對答案

❶ off the list price

❷ quantity (or volume) discount program

❸ absolute lowest

❹ further；sales targets

說說文，解解字

要說折扣從哪兒算下來，或是哪個產品、哪張帳單會提供折扣，就可用「xx% discount off + 名詞」來說，像若要說首購可享25%的折扣，就可寫「We could offer you a 25% discount off your first purchase.」，若要告訴客戶現在訂購可享定價10%折扣的特惠，就請這麼說：「The following products are currently on special offer with a 10% discount off the catalog price.」，若要來個大放送，讓客戶帳單折個25%，那就請大聲告訴對方「You will get 25% discount off your bill.」！

數量折扣說quantity discount或volume discount都可，那quantity與volume完全一樣嗎？也不是這樣說哩！我們來看看搭配詞，在quantity的部分，若要說用掉的材料的量，會說the quantity of materials used，要指特定的一個量，會說consumed in large quantities 或是take a small quantity of drugs。Quantity本身也有大量的意思， 所以要說量大時就會便宜些，可說「It is cheaper when bought in quantity.」。而說到volume，車流量是volume of traffic，貿易總額是total volume of trade，說工作量大會說huge volume of work…對這兩

個字的差別有感覺了嗎？正是quantity多指具象，volume多指抽象！順道一提，volume還有別的字義，包括體積、容積、容量、音量、（書）冊與卷，這些也就只能說volume了，沒quantity的事！

第4題的further一字，常會拿來和farther比一比，further與farther都可用來指稱遠些的距離，但我們只能用further discounts，而不用farther，原因就是further還可表抽象概念，所以要說進一步、額外的折扣，就只能用further了。

搞定報價

1-3 交貨條件

Allen: Allen Fang speaking. How may I help you?

Mary: Hi Allen. This is Mary Lopes from Sunrise Company in Germany. I was given your number by Elizabeth Rosenberg. She's a customer of yours.

Allen: That's right. What can I do for you, Mary?

Mary: Well, she told me that you have decorative glassware of quality. I'd like to know more about your products, like what models do you have and which are the best-selling items?

Allen: I see. We have various models with different colors and sizes of handmade decorative glassware. All of them are with best quality and the lowest price. You can visit our website to know more about our products.

Mary: OK. Oh, one more thing before I forget. Which shipping term do you quote?

Allen: We offer Ex-Works quotes unless requested otherwise.

Mary: Can you quote to us FOB shipping terms instead? We prefer that all related costs to get the goods onto your airport are to be covered by your side.

Allen: Not a problem. We can also arrange FOB shipments. So

please review the product information on our website and then tell me which models are of your interest. Or you could give me your Email address and I'll send you first the product information of our best-selling items. I think you'll find them extremely appealing and attractive!

Mary: I can't wait to see them! My Email address is Marylopes@sunrise.de. I'll keep an eye out for it!

Allen: Thanks for your interests in our products!

背景說明	
人物	Allen：原廠業務代表 Mary：客戶
對話主題	價格條件： 工廠出廠價、FOB價

譯文

艾倫：我是方艾倫，請問有什麼事嗎？

瑪莉：嗨，我是德國日昇公司的瑪莉．洛佩茲。伊莉莎白．羅森柏格給了我您的電話號碼，她是您的一位客戶。

艾倫：沒錯，有什麼需要我幫忙的嗎？

瑪莉：嗯，她告訴我您們有品質很好的裝飾玻璃器皿，我想要多瞭解一些您們的產品，像是有哪些類型？哪幾款銷售得最好呢？

艾倫：我瞭解了，我們有好幾種不同類型的手工裝飾玻璃器皿，各有不同的顏色與尺寸可供您選擇，每一項都有最高的品質，最划算的價格。您可以到我們的網站看看，就可對我們的產品有多一點的瞭解。

瑪莉： 好的。喔！還有一件事，趁我現在記得時先問一下，您們所報的會是什麼樣的交貨條件呢？

艾倫： 除非另外有指定，要不然我們報的都是工廠出廠價。

瑪莉： 您可以報FOB條件的價格給我們嗎？我們希望貨到您機場的所有相關成本都由您這邊來負擔。

艾倫： 沒問題，我們也可以安排以FOB條件出貨。所以就請您到我們的網站看看產品訊息，再告訴我您有興趣的是哪幾款，或者，您也可以告訴我您的Email地址，我可以先寄一些我們銷量最好的幾款產品的資料給您，我想您會覺得這些產品都很有魅力、很吸引人呢！

瑪莉： 我等不及了呢！我的Email地址是Marylopes@sunrise.de，我會留意的。

艾倫： 謝謝您對我們的產品有興趣！

例一 | Ex-Works工廠出貨價

例句 Unless otherwise agreed between the parties, the terms of delivery are Ex-Works, Taipei, Taiwan.

譯文 除非雙方有其他協議，否則議定的交貨條件即為臺灣臺北的工廠出廠價。

例二 | Ex-Works工廠出貨價，可安排配合的貨運公司

例句 Currently, the prices we are offering to customers are Ex-Works. For that reason, we need to know if you have any arrangement with a shipping company. If needed, we will be pleased to help you out with the shipment. But, since you are a distributor, I guess you can have cheaper prices for shipping services.

譯文 我們目前所報給客戶的都是工廠出貨價，因此，我們必須知道是否您有可配合的貨運公司，若有需要，我們很樂意提供協助，安排出貨，不過，既然您是經銷商，我想您應可拿到更划算的出貨費率。

例三｜FOB船上交貨價，運費另計

例句 The price for PC-101 is 230 USD per vial FOB Kaohsiung, Taiwan. Freight is additional and depends on the quantity ordered.

譯文 產品PC-101的價格為每瓶美金230元，此為臺灣高雄的船上交貨價，運費會另計，金額視訂購數量而定。

例四｜FCA運送人交貨價，配合FedEx出貨

例句 Consolidated weekly shipment to your company will be made by us via FedEx International Priority with shipping charges made to your company's FedEx account. All shipments made by us are FOB Kaohsiung, Taiwan.

譯文 每星期所統整的出貨會是透過聯邦快遞國際優先服務型的方式出給您的公司，運費會計入您公司的聯邦快遞帳號，而我們安排的所有出貨皆是以在臺灣高雄的運送人交貨價為條件。

例五｜FOB船上交貨價，說明成本與風險歸屬

例句 For FOB shipping terms, we are responsible for the costs and risks of getting the goods on board the ship. Once the goods are loaded on the ship, the costs and risks are then transferred to your side. So this means you will pay the ocean freight from the port of origin, plus all of the other transport charges and customs clearance cost etc.

譯文 對於船上交貨價的交貨條件，我們會負擔貨物送至船上的成本與風險，待貨物裝船後，之後的成本與風險就轉由您們負擔，也就是說您須負擔從出口港口開始計算的海運運費，另外再加上所有其他的運輸費、報關費等費用。

關鍵字急救站—terms

這一個單元說到了很多的shipping terms／交貨條件，在這裡我們就來好好地看一看這個我們在國貿出貨文件上、合約上一定會看到的字：「terms」……咦？這個字我們算熟的了，那……那一定要用複數、要加s嗎？是的，當它當作「條件」來解釋時，它就是要寫為複數型態。先來看看它的意思：「the conditions of a legal, business, or financial agreement that the people making it accept」，說的是與法律、商業、財務有關的合約與協定條件即為terms，所以，這一個單元說的是shipping terms／交貨條件，而談到付款條件時，就會說 payment terms或是 terms of payment，此外，在原廠出具的每一份Sales Order／銷售訂單、Order Confirmation／訂貨確認單、Invoice／發票，裡頭一定都會寫到shipping terms及payment terms這兩種條件。

再來，在外貿業務中，我們也常看到這個詞組：「terms and conditions」出現在e-mail或原廠出具的許多文件裡。原廠常會發來與訂單、出貨、付款相關的terms and conditions，在客戶看過並接受後，原廠才會與客戶繼續下一步的交易動作。像是在原廠的網站中，常會見到這一句：「Click here for terms and conditions.」，就是請你點入連結，看看原廠的相關條件規定。另外，像是在FedEx的每一封出貨通知e-mail裡，最下方都會列出一些注意事項，裡頭就有寫到：「Please see FedEx Service Guide for terms and conditions of service, including Fedex Money-Back Guarantee.」，說的是請你參閱FedEx服務指南上所列之服務條件，其中也包括了他們的退款保證規定。

若原廠要說明這些條件會有變更，但並不會特別一一通知客戶，則會寫上這樣的注意項目：「These terms and conditions (“Terms and Conditions”) may be updated by us from time to time without notice to you.」。最後，若要問對方是否接受這些條件，就會這麼問：「Do you agree to these terms and conditions?」，簡單吧？！而簡單又複雜其實就是terms的本質啦！

「terms」這個字簡單又常見，而terms裡頭的條件與條文可就不是三言兩語說得完、道得盡的，要是你卯起來給將原廠的terms逐條看過，那你就會成為「Terms達人」啦！

術語直達車，專業補給站

説到交貨條件，裡頭的術語可多著呢！若你喜歡看頭字語來猜全稱，那你在Incoterms的這些交貨條件中就會找到許多的樂趣呢！先來看Incoterms的全稱，Incoterms stands for International Commercial Terms，它簡稱為國貿條規，是由International Chamber of Commerce（ICC）／國際商會所制定。

在2010年新版Incoterms中，共列有11種的交貨條件及規範，説明了國際貿易中買賣雙方所應負的責任、成本與風險，我們在這裡就來説説幾種常見常用的交貨條件囉！

EXW（Ex-Works）｜工廠交貨價

「Ex-」是「從……」、「在……交貨」的意思，「Works」指的是「工廠」，所以Ex-Works這種交貨條件，指的就是工廠交貨價。若賣方報價的交貨條件為EXW，則賣方一點兒也不用負擔任何出貨的成本，也不用承擔出貨途中可能發生的任何風險，只要把貨物準備好，等著買方安排的貨運公司來賣方這兒取貨即可，而將貨物裝上運送人車上的責任，也是買方的事，賣方在一旁納涼即可。

所以，在這樣的條件裡，買方就得自理所有出貨相關事宜，找配合的貨運公司，打理好保險。對於出口貿易來説，在這樣的交貨條件下，若貨物出口還得要出口國家主管機關出具許可證或其他的證明文件，雖賣方可提供協助，但就一樣得由買方自己聯繫並自負費用與風險了，所以，若須出口證明文件，Ex-Works可能就不是對買方有利且方便的交貨條件了。

FOB（Free On Board）｜船上交貨價

「Free」指的是「解除賣方責任」，「On Board」是指「在船上」，所以Free On Board的條件表示的就是賣方對貨物的責任就到上船為止。賣方所報的這種交貨條件，代表的是他們須負擔將貨物送上船之前的所有成本與風險，等到貨物一上了船，成本與風險就都轉由買方來承接了，舉凡從出口港到買方收到貨之前所發生的任何運費、保險費、報關費、稅負等成本的帳單，一律都是開給買方！

在FOB的交貨條件下，辦理貨物出口手續的工作還是在賣方，買方不用遠從海外插手辦理，所以對買方來說，FOB的交貨條件就比EXW更讓人順心愉快，因此也成為了國貿上買賣雙方常用的交貨條件了。

FCA（Free Carrier）｜運送人交貨價

剛說過「Free」是「解除賣方責任」的意思，而「Carrier」是指「運送人、運輸業者」，所以Free Carrier的條件表示的就是賣方的責任就到將貨物交給買方所指定的運送人為止。在此交貨條件下，買方須選定運送人，將其連絡資料與指定的交貨時間告訴賣方，賣方須負責貨物送到指定地點、交給送貨人之前的費用與風險，等到貨物交到送貨人手上後，即可解除責任。此外，交貨地點的不同，在責任歸屬上也會有所差別，若是約定在賣方的所在處所交貨，則賣方須負責將貨物裝上運送人的車輛上，若是所約定的地點是在賣方所在處所以外的任何地方，則賣方的責任是到將貨送到指定地點就中止，買方要負責將貨物從賣方車輛上卸下與裝上自己的車輛。

FCA的條件與FOB有什麼不同呢？那個…FOB既然叫作船上交貨價，所以它管海運，FCA交貨給運送人，所以陸海空都可以管！確實字面上是這麼說的，但在國貿實務上，就算買賣雙方談論的都是空運，還是會把FOB說得嚇嚇叫！若說到費用負擔上的不同，那FCA與FOB就有點不同了，賣方報FOB價格，會比在賣方所在處所外地點交貨的FCA，多負擔於出口港口將貨物卸下以及再將貨物裝上船的費用。

CFR（CNF／Cost and Freight）｜運費內含價

「CFR」的「FR」取自「Freight」的頭兩個字母，再請你往上看回FCA……它的「CA」也是取自「Carrier」的頭兩個字母……所以你有沒有看出一點頭字語的一個取名方式了？那就是當字母數未達想要的數目時，就請取用一個字的頭兩個字母吧！在CFR的條件下，賣方的報價除了產品成本之外，還需包括將貨物一路送到目的地港口的運費，但不包含保險費。

CIF（Cost, Insurance and Freight）｜運保費內含價

這種交貨條件比CFR多了一個Insurance，所以賣方的報價除了產品成本之外，還須包括貨物一路送到目的地港口的運費及保險費。

DAP（Delivered at Place）｜目的地交貨價

這個條件以前叫作「Delivered Duties Unpaid」，（未完稅交貨價）賣方要承擔的責任從其工廠出貨開始，直到將貨物交到目的地（通常是買方的所在處所）才能終止，不過，這個交貨價並不包含海關清關的費用以及相關的稅負。

DDP（Delivered Duties Paid）｜完稅後交貨價

此條件比起DAP更完滿了，DAP不含清關費用與稅負，而DDP則是通包，從出貨地到目的地整程所有的費用與風險，通通都由賣方負責。因此，買方在出貨安排上，確實可以都在一旁納涼，不用插手。在DDP的條件下，賣方責任最大，買方責任最小，而在Ex-Works條件下則是恰恰相反，賣方責任最小，買方責任最大。

除了上述這七種交貨條件之外，另外還有四種條件：

- FAS（Free Alongside Ship／船邊交貨價）
- DAT（Delivered at Terminal／終點站交貨價）
- CPT（Carriage Paid to／運費付訖交貨價）
- CIP（Carriage and Insurance Paid to／運保費付訖交貨價）

若你哪天碰到了這四種條件，就請你再去查看它們的說明囉。最後，針對前面說到的七種交貨條件，我們就來細看一下根據Incoterms 2010所整理出的買賣雙方責任歸屬表囉！

Incoterms 2010所整理之買賣雙方責任歸屬表：

條款	EXW	FCA	FOB	CFR	CIF	DAP	DDP
Loading on truck (carrier) **卡車裝貨 (貨運公司)**	買方	賣方	賣方	賣方	賣方	賣方	賣方
Export-Customs declaration **出口海關申報**	買方	賣方	賣方	賣方	賣方	賣方	賣方

條款	EXW	FCA	FOB	CFR	CIF	DAP	DDP
Carriage to port of export 運貨至出口港口	買方	賣方	賣方	賣方	賣方	賣方	賣方
Unloading of truck in port of export 出口港口卡車卸貨	買方	買方	賣方	賣方	賣方	賣方	賣方
Loading charges in port of export 出口港口裝運費用	買方	買方	賣方	賣方	賣方	賣方	賣方
Carriage to port of import 運貨至進口港口	買方	買方	買方	賣方	賣方	賣方	賣方
Unloading charges in port of import 進口港口卸貨費用	買方	買方	買方	買方	買方	賣方	賣方
Loading on truck in port of import 進口港口卡車裝貨	買方	買方	買方	買方	買方	賣方	賣方
Carriage to place of destination 運貨至目的地	買方	買方	買方	買方	買方	賣方	賣方
Insurance 保險	買方	買方	買方	買方	賣方	賣方	賣方
Import customs clearance 進口通關	買方	買方	買方	買方	買方	買方	賣方
Import taxes 進口稅	買方	買方	買方	買方	買方	買方	賣方

Show Time! 換你上場！

❶ 我們在這一個單元認真地看過了這麼些個交貨條件，當然要來好好試試我們自己的能耐，看我們是否都記得了它們的英文全稱？是否也可順口叫出它們的中文呢？好了，不囉嗦，動手填空吧！

交貨條件頭字語	英文全稱	中文
EXW		
FOB		
FCA		
CFR		
CIF		
DAP		
DDP		

❷ 請寫出Incoterms的全稱：

________ ________ ________

❸ 在上述七項交貨條件中，請問有哪三項不含運費：

1）________

2）________

3）________

來對對答案

❶ 請你往前翻翻查查囉！

❷ International Commercial Terms

❸ 1）EXW　2）FOB　3）FCA

搞定報價

1-4 計算報價

Daniel: Hi, Jenny. It's Daniel from Vision Chemical.

Jenny: Hello, Daniel. I've been meaning to get back to you about your offer.

Daniel: Yeah, that's the reason I called. What do you think of our offer?

Jenny: You offered us your EX-works price. We'd also like to know your shipping and handling fee.

Daniel: The shipping fee is 100 USD. If you prefer, you may provide your FedEx, DHL or UPS account for shipping and we will not charge a shipping fee. In addition, since your inquired product needs to be shipped with dry ice, a 90 USD dry ice packaging and handling fee will be added to the Invoice.

Jenny: Could you give us the package's gross weight and dimensions for us to check the freight with our courier agents?

Daniel: No problem. The box will contain 8kgs weight and the dimension are 35x35x30 cm.

Jenny: Thanks. After price comparison, I'll confirm with you whether we want to use your own courier account to ship for us.

Daniel: Alright. Do you need any other information?

Jenny: Oh. You do remind me. What's the documentation fee included in your quote?

Daniel: The product is a hazardous material. It's controlled in our country, and so it requires an export permit for international shipment.

Jenny: I see. I'll calculate and check with my supervisor and then give you a call tomorrow about our order.

背景說明	
人物	Daniel：原廠客服代表 Jenny：客戶
對話主題	價格條件與相關費用： 工廠出廠價、運費、處理費、包裝費、文件費

譯文

丹尼爾：嗨，珍妮，我是遠見化學的丹尼爾。

珍　妮：丹尼爾，您好，我一直想回電給您，談談您的報價呢！

丹尼爾：那也是我打電話找您的原因呢！您覺得我們的報價如何呢？

珍　妮：您報給我們的是工廠出貨價，我們也想知道一下運費和處理費的金額。

丹尼爾：運費是美金100元，您要的話，也可以走您的FedEx、DHL或是UPS的帳號出貨，這樣的話我們就不用跟您收運費了。另外，因為您所詢的產品得用乾冰出貨，所以美金90元的乾冰包裝與處理費會再加進發票裡。

珍　妮：您能否給我們出貨包裝的毛重與尺寸呢？我們想跟我們快遞公司報的運費比較一下。

丹尼爾：沒有問題，箱子會有8公斤重，尺寸是35x35x30公分。

珍　妮：謝謝。比價後我會再跟您確認是否要走您們的快遞來出貨。

丹尼爾：好的，您還有什麼要問的嗎？

珍　妮：喔！您真的提醒了我！您報價上有列文件費，那是什麼呢？

丹尼爾：這產品屬於危險物質，在我們國家有管制，所以要申請出口許可。

珍　妮：瞭解。我會算一算，也會跟我主管談談這個訂單，明天我再回您個電話。

說三道四，換句話試試

例一｜請提供FedEx帳號供辦理出貨

例句　I have attached a quote for one PK101 for your information. This includes shipping and handling fees. If you prefer, you may provide us with your FedEx account number so that these charges will be billed directly to you. Otherwise, we will use our own account, and the delivery charges on the quote will apply.

譯文 為了讓您瞭解，我在此附上一組PK101的報價如附，此價格包括運費與處理費，您要的話，也可提供您的FedEx帳號給我們，那這個運費就會直接加到您的帳上，否則，我們就會用我們自己的帳戶來出貨，運費就會是報價單上的金額。

例二 | 銀行手續費及特殊包裝費另計

例句 The distributor price is 1175 USD per 1 mg, ex-works Taiwan. There will also be the standard 25 USD fee to cover bank charges from international wire transfer. This item ships on dry ice at -20°C. There will be a handling charge for the dry ice and special shipping box of 56 USD.

譯文 每1毫克的經銷商單價為美金1175元，此為臺灣工廠的出廠價。對於國際匯款，我們還會加收美金25元的銀行手續費標準收費。這個品項須維持在-20°C的溫度下，要用乾冰出貨，所以還會有乾冰特殊出貨箱的處理費美金56元。

例三 | 銀行手續費另計

例句 Regarding your method of payment, we can accept credit cards, or a bank wire transfer. If you choose to pay via bank wire transfer, an additional 25 USD charge should be added to the total on your quote.

譯文 在付款方式這部分，我們可接受信用卡或是銀行匯款，若您選擇銀行匯款，則在給您的報價上還會另外加上美金25元的額外費用。

例四｜信用卡付款手續費另計

例句 You can pay by wire transfer or check. Payment by credit card is also acceptable, and we may surcharge the 3% processing fee of credit card transactions.

譯文 您可以用匯款或支票來付款，我們也可接受信用卡，信用卡交易會另收3%的信用卡處理費。

例五｜透過經銷商購買可省去的費用

例句 While direct sales pricing is available on our website, these will be different from the price through our distributors. In many cases, ordering through these distributors will avoid most of the following additional fees that may apply to your direct order:

Shipping and handling fee: 95 USD

Export documentation fee: 20 USD

Wire transfer fee: 20 USD

譯文 我們網站上有列出直購價，這些價格會與您透過我們經銷商購買的不同，透過這些經銷商訂貨大多都能省去下列費用：

運費與處理費：美金95元

出口文件費：美金95元

匯款手續費：美金20元

關鍵字急救站—fee

「費費」大集合！

話說除了產品成本之外，還有包裝費、文件費、運費、銀行手續費、信用卡手續費等費用，這麼多「費」，是不是一「fee」到底呢？說完「fee」之後，還有沒有別的用法呢？我們就在這裡好好整理一下「費費」事，一定讓你費點小力後，以後下筆寫「費」再也不費神啦！

我們會看到的「費用」說法包括有fee、charge、surcharge、fare、expense和cost，在這兒就先來看看它們的英英解釋囉：

- fee: a charge fixed by law or by an institution for certain privileges or services.（法定或機構所訂之定額費用，以取得某特權或服務）
- charge: an amount of money that you have to pay especially when you visit a place or when someone does something for you.（特別是參觀某地或請某人為你做事時所須支付的金額）
- surcharge: an additional amount of money that you must pay for something over the usual price（為取得某產品或服務所支付之一般收費之外的額外金額）
- fare: the money that you pay for a journey（為旅程所支付的金額）
- expense: an amount of money that you spend in order to buy or do something（為了購買某物或能夠做某事所花費的金額）

➡ cost: the amount of money that is needed in order to buy, pay for, or do something （為了購買某物或能夠做某事所需的金額）

看了這些字詞的解釋後，我們可明確地將surcharge與fare從這些費用中獨立出來。surcharge是額外加計的費用，例如「All payments are due in US dollars or add 10% surcharge for payments in other currencies.」，說的是應以美金付款，若是以其他貨幣支付，則要加收10％的額外費用，又如「Payment by credit card (Visa or MasterCard only) is also acceptable, and we may surcharge the 3% processing fee of credit card transactions.」，指的是也可接受信用卡付款，但會加收3%的信用卡交易處理費。至於fare呢，在應用上可單純了，是指跟交通有關的費用，如「We will afford all of your accommodation and transportation fare.」，說的就是會為來訪貴賓支付所有的住宿與交通費用，另外說到搭乘交通工具也都會說到這些fare：air/bus/train/taxi fare。

我們再來瞧瞧fee與charge這兩個在字義上有類似之處的費用了。兩字較量之後，在專業、執業這點上，fee略勝一籌，像是律師或醫生所收取的費用，多是用fee，像是「license fee」這類要求出具證照或特許權所收的費用。而charge相較之下就沒這麼正式了，通常也會用於指稱為獲取服務所支付的小額金錢。

在實務上，除了license fee由fee專攬之外，其他在產品價格之外收取的金額其實都是小額，所以會看得到fee與charge競相出現，指稱同一種費用。請您瞧瞧下列這些fee與charge皆有使用的費用名目囉：

- 運費：shipping fee／charge
- 處理費：handling fee／charge
- 運費與處理費：shipping and handling fee／charge
- 包裝費：packaging fee／charge
- 乾冰費：dry ice fee／charge

所以呢！說額外費用請用surcharge，說搭乘交通工具的費用請說fare，除此之外的與訂單處理相關的包裝與出貨等費用，就請fee與charge隨心所欲地使用囉！

那最後所列的「expense」又該怎麼用呢？我們也看過「packing expense」這類的用法呀？沒錯，這樣說一點兒也沒錯，只不過它的立論出發點是從原廠的支出面來看。從原廠的角度來說，包裝費用的支出就是packing expense，處理手續所衍生的費用支出就是handling expense，而這些費用都是原廠的「cost」／成本了。要將這些支出費用或成本轉嫁給客戶，跟客戶收取，那就得要正其名而成為賣方所報、買方所須支付的fee、charge、surcharge或fare囉！

術語直達車，專業補給站

廠商生產了產品，定了價，也定了經銷商價格之後，可還有好些個行政作業的費用得要理一理，而且在客戶詢價時，就應該一併提供給客戶，才不會原先沒說，後來才補，也不會在說了一項後，最後又來記回馬槍，再補一筆！到底有哪些行政費用呢？我們在此一起來看看將貨物送去遠渡重洋會產生哪些常見的出口費用囉！

一、包裝費：Packaging or Packing fee／charge

包裝這詞說Packaging或Packing都可，但若要細分來看，packaging是內包裝，是為銷售、行銷考量而做出的包裝，而packing指的是為了運送、出貨而有的外包裝。若是出貨產品有溫度或其他要求，需特殊處理，則會再另行加計特殊包裝費，如須以乾冰出貨，則會加收乾冰費：dry ice fee／charge。

二、處理費：Handling fee／charge

原廠處理訂單皆會涉及行政文書、倉儲等作業，因此原廠會要跟客戶收取這部分的成本，也就是處理費（handling fee／charge）。許多原廠會將處理費與運費（shipping fee／charge）合併在一起，所以會出現這麼一條費用條目：shipping and handling fee／charge，或是簡寫為「S&H」。

三、運費：Shipping fee／freight

若原廠所報價格是以像是CIF這種含運費的交貨條件，原廠就需要計算好運費，若是以FOB為貿易條件，原廠在報價中並不需要計入運費，但客戶也有可能詢問運費的大概金額。要算運費，可分成兩段來看：

- 國內運費：Inland shipping fee／charge
  將貨品運送到出口地點所需的運費，如陸路的火車或卡車運費以及搬運費用等，即屬於國內運費。
- 國際運費：International shipping fee／freight
  貨品從機場或碼頭出口所需支付的作業費用，即是國際部分的運費。以空運來說，會產生的費用除了空運運費之外，還會有這些額外加計的費用，例如兵險費（war risk surcharge）、燃油費（fuel surcharge）以及倉租費（storage fee）等。

四、保險：Insurance

有些交貨條件須由原廠處理保險事宜，向保險公司投保，如在CIF與CIP條件之下，原廠須按照約定的保險險別與金額，辦理投保，支付保險費。投保金額若無特別要求，則會按照CIF或CIP的價格加成10%來辦理。若買方要求增加加成比例，則增加的保險費將再加計上去，由買方負擔。若是以像是FOB、CFR或FCA條件成交的訂單，則出口商僅須提供相關資料，供客戶自行辦理保險。

五、出口稅捐：Export taxes and duties

出口稅捐包括了出口關稅、商港服務費、推廣貿易服務費等。

六、文件費：Documentation fee

貨品若是在出口國家有出口管制，或是屬於法定檢驗商品，則須在申請了出口許可或相關的檢驗與證明文件後，方可出口，這些文件的費用都可稱為documentation fee，若是針對申請出口許可者，當然也可直接點名說它是export permit／license fee。

七、銀行手續費：Bank fee

國際匯款由於跨國銀行間辦理程序的規定，會收取一筆手續費，有些是匯款人在辦理匯款手續時額外收取，有的則是直接從匯款金額中扣抵。為免實際匯入帳戶的金額短少，原廠會要求匯款人支付所產生的手續費，確保全額到行。

另外，客戶以信用卡支付貨款時，若有信用卡扣款手續費需支付，則原廠亦應在客戶確定付款方式前，告知客戶此手續費的計費比例。

最後，就讓我來順一次從國內銷售價格到出口所報的「CIP運保費付訖交貨價」的費用，看這一路會加計什麼樣的費用，也複習一下幾種交貨條件的所屬定位囉！

國內銷售價格
出口包裝
出貨前檢查處理
出口文件（如產地證明）
銀行費用
EXW工廠出廠條件報價
運貨至港口、機場
FOB船上交貨價／FCA運送人交貨價
海運運費、空運費用
CPT運費付訖交貨價
保險
CIP運保費付訖交貨價

Show Time! 換你上場！

❶ 您報給我們的是工廠出貨價，我們也想知道一下運費和處理費的金額。

You offered to us your ________ price. We'd also like to know your ________ ________ ________ ________.

❷ 您能否給我們出貨包裝的毛重與尺寸，讓我們比較一下運費。

Could you give us the package's ________ ________ and ________ for us to check the freight ?

❸ 因為您所詢的產品得用乾冰出貨，所以美金90元的乾冰包裝與處理費會再加進發票裡。

Since your inquired product needs to be shipped on dry ice, a 90 USD dry ice ________ ________ ________ ________ will be added to the Invoice.

❹ 請寫出下列費用的英文：

➡ 運費：________________________________

➡ 處理費：________________________________

➡ 運費與處理費：________________________________

➡ 包裝費：________________________________

➡ 乾冰費：________________________________

來對對答案

❶ EX-works；shipping and handling fee

❷ gross weight；dimensions

❸ packaging and handling fee

❹ shipping fee / charge
handling fee / charge
shipping and handling fee / charge
packaging fee / charge
dry ice fee / charge

Chapter 1

搞定報價

1-5 經銷商的價格結構

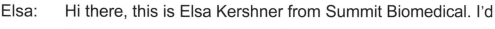

Elsa: Hi there, this is Elsa Kershner from Summit Biomedical. I'd like to speak to John Chen, please.

John: Hi, Elsa. Nice to hear from you. How've you been?

Elsa: Not bad. I'm calling to ask for your assistance by giving us a larger discount for AML-0202 model instrument.

John: You've got 40% distributor discount for this instrument. How big is the discount that you exactly want?

Elsa: In order to compete, we hope you could approve a 50% discount.

John: That's really a large discount... Hmm... We're able to lower our price to you, but we expect you to adjust your profit margin, too.

Elsa: Sure. Our previous offer to customer is 5,500 USD which is already close to our landed cost and we got almost zero profit margin. But the competitor goes for an even lower price, and it's impossible for us to do this business without your support of a larger discount.

John: I see... In order to win this project, I could approve another 5% discount.

Elsa: Thanks for your support! I'll keep you informed of any progress.

背景說明	
人物	Elsa：經銷商業務代表 John：原廠業務經理
對話主題	要求折扣相挺： 原廠能否折扣給多一點？經銷商能否利潤調低一點？

艾爾莎：嗨，我是高峰生醫的艾爾莎・科許納，麻煩請找陳約翰。

約　翰：嗨，艾爾莎，很高興接到您的電話，您好嗎？

艾爾莎：還不錯。我打來是要問問您能否幫忙在AML-0202型的儀器價格上再多給些折扣。

約　翰：您們在這個儀器上的價格已有六折的經銷商價，您想要的是多大的折扣呢？

艾爾莎：為了競爭所需，我們希望能有五折的價格。

約　翰：那真的是個大折扣耶…嗯…我們是有辦法再降些價格，但也希望您們在獲利率上也能有所調整。

艾爾莎：當然。我們先前報給客戶的價格是美金5,500元，這個價格已經快跟我們的落地成本差不多了，幾乎是零利潤，結果競爭者報的價格居然更低，因此，若沒有您的協助、多給些折扣，我們就根本沒辦法做這個生意了。

約　翰：瞭解…為了要拿下這個案子，我同意再多給您5%的折扣。

艾爾莎：謝謝支持！若有什麼新進展，我就會馬上跟您回報。

說三道四，換句話試試

例一｜原廠給了更多的折扣，請經銷商也少賺一點哩！

例句　Since we are giving a higher discount to compete, we expect you to lower your margin in order to provide a competitive offer to the customer.

譯文　既然我們為了競爭而給了較大的折扣，我們希望你們也能降低你們的利潤，以能提供個有競爭力的價格給客戶。

例二｜原廠利潤也少，但為表善意，若訂單加碼，折扣也可加碼，再降5%！

例句　We offer distributor prices to your company, with a limited margin for us. For the discussed product, its distributor discount is already 30%. However, in good faith, we can offer another 5% discount for a minimum order of 3 kits. Kindly please let us know if you would like us to provide you with an official Quotation or Proforma Invoice.

譯文　我們是報經銷商價格給你們公司，在這樣的價格上，我們的利潤有限。關於討論到的這項產品，經銷商折扣已經有30%，但為表善意，若你們最少能訂到3組，則我們能再多給5%的折扣。還請告知是否要我們提供正式的報價單或形式發票給你們。

例三｜原廠對量大的訂單可再降價，也請經銷商降低利潤。

例句　We have reviewed our pricing and we are able to lower the price to you. In turn, we expect you to adjust your margin so that you can obtain the customer's order. Attached is our revised quotation. This pricing is valid only on the total order of 1 gram x 10 vials.

譯文 審視了我們的定價之後，我們是有辦法降一些價格給你們，相對地，我們希望你們也能調整利潤，這樣你們也才能拿下客戶的訂單。在此附上我們修改的報價，此報價只適用於總量有1克×10瓶的訂單。

例四｜原廠已給了經銷商折扣，恕難再降！

例句 Unfortunately, 1,440 USD distributor price is as low as we can go on that product. We just don't have much margin to work with when it comes to our distributor prices. Sorry I can't help out.

譯文 抱歉，經銷價美金1,440元已是我們所能給的最低價，在經銷價這個價格水準上，我們其實沒太多利潤。抱歉沒能幫上忙。

關鍵字急救站

說到利潤，一般人先想到的會是「profit」，不過也有看過國外寫「margin」……那它跟profit的意思一樣嗎？在經濟學上倒是說到很多的經濟量時，都有「marginal」／「邊際的」這個字，像是marginal cost／邊際成本、marginal revenue／邊際收益、marginal profit／邊際利潤，說的都是增加一個單位時所增加的總量，那profit margin說的是利潤邊際嗎？這詞兒聽起來邏輯不太好懂耶！

我們先來看看profit margin的英英解釋：「the difference between

how much money you get when you sell something and how much it costs you to buy or make it」，profit margin說的就是一件產品之賣價與買價／成本的差額，而寫到profit margin的數值時，就會出現百分比的表示法，所指的也就是獲利率了，正如此段文句所說：「Profit margin is profit divided by sales. Profit margins can provide a comparison between companies in the same industry, and can help identify trends in the numbers for a company from year to year.」，此處說到profit margin就是以利潤除以銷售額，有了獲利率，就可看出一個產業的整體經營狀況，也可讓同產業裡的各個公司有個評估、比較的指標，單就一家公司幾年來獲利率的變化來看，也可評估這家公司營運模式的成長狀況與所面臨的壓力。

而在計算報價時，除了看到margin之外，有時也會看到「markup」／加成這個字，而從markup計算售價跟從「margin」來算，可是兩碼子的事，基準一點兒也不同呢。所謂的margin是從售價／final selling price上來算的一個比率，即獲利率，而markup percentage則是「the percentage of the cost price you add on to get the selling price」，這種方式所訂出的售價，是從成本加成百分比之後而得。Markup與margin都是高出成本的部分，而要確實地一眼看穿獲利能力，可就非margin不可了！

此外，在計算能給經銷商或客戶多少折扣時，只要知道profit margin為多少百分比，當經銷商或客戶對著你大喊著「給我折扣～」時，你就能知道給了折扣後還剩下多少獲利率，給多少折扣才不會跟自己的利潤過不去或變成做白工…要做白工可以，但也要有夠份量的市場策略因素考量才行呢！

最後就讓我們來實地算算數字，將這些個cost、markup、margin、revenue好好理一理！

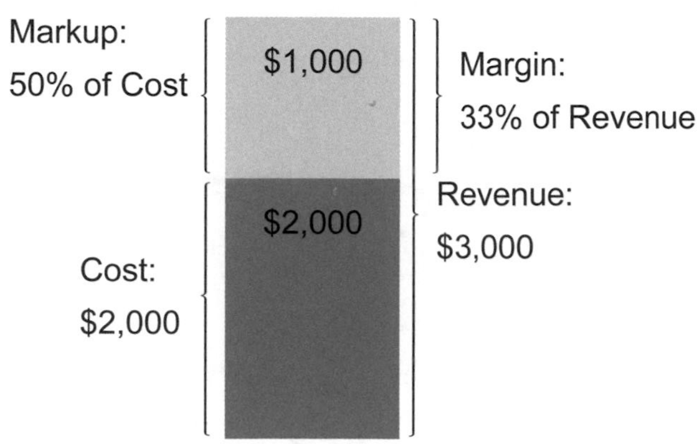

圖中的這些markup & margin的百分比數值，你都算對了嗎？這一個單元最後還有個數學練習的情境題，還可讓你盡興算翻天呢！

術語直達車，專業補給站

當經銷商跟原廠要特價時，尤其是要大折扣時，有時原廠會想：「那你經銷商是加了多少利潤之後才報給客戶呢？」，或是「我們原廠給了特價折扣，這些折扣有沒有都轉給了客戶呢？還是這部分的差額其實是成了經銷商的利潤？」當原廠有這樣的疑慮，那原廠就會要開口問經銷商給客戶的報價金額：「What's your offer to customer?」，或是直接問經銷商的價格結構了：「Please let us know your price structure.」。

經銷商的價格結構有哪幾個層次呢？原廠方面可控制到的有原廠自己所定的list price／定價，以及給予經銷商特有折扣的distributor price／經銷商價格，之後所言及的價格可就是經銷商方面所處理與運作的了。貨品從原廠出貨、最後到達經銷商手上的所有成本總額，稱為落地成本，而經銷

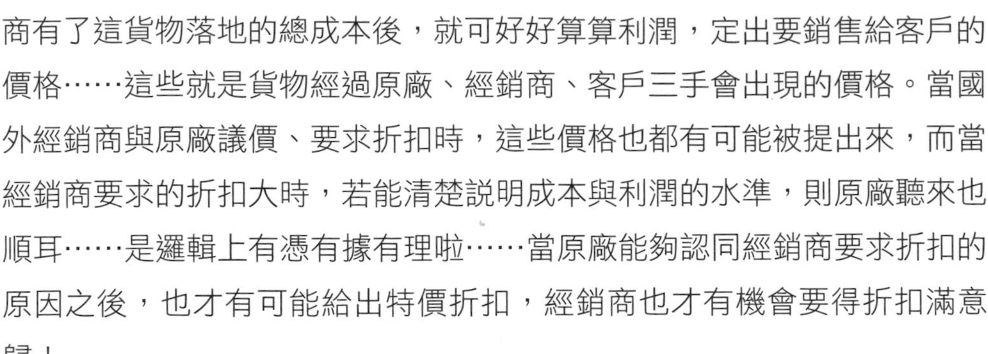

商有了這貨物落地的總成本後，就可好好算算利潤，定出要銷售給客戶的價格……這些就是貨物經過原廠、經銷商、客戶三手會出現的價格。當國外經銷商與原廠議價、要求折扣時，這些價格也都有可能被提出來，而當經銷商要求的折扣大時，若能清楚説明成本與利潤的水準，則原廠聽來也順耳……是邏輯上有憑有據有理啦……當原廠能夠認同經銷商要求折扣的原因之後，也才有可能給出特價折扣，經銷商也才有機會要得折扣滿意歸！

接著，我們來用圖説來清楚説一説，來從原廠報價一路順到給客戶的報價囉：

List price定價

Distributor price經銷商價格
Discounted price（Disc. price）折扣後價格
Transfer price原廠報給經銷商的售價

Landed price落地成本
貨品到達經銷商處之前的所有成本：
產品成本＋運費＋報關費＋稅負等成本

End user price最終使用者／客戶價格
（總成本＋利潤）

List price、Distributor price、Transfer price這幾種價格在前面單元有介紹過了，我們在這兒就來看看什麼叫做「landed cost」囉！

landed cost:

The total cost of a landed shipment includes purchase price, freight, insurance, and other costs up to the port of destination. In some instances, it may also include the custom duties and other taxes levied on the shipment.

這段文字説明了落地成本指的是一項產品從空中、從海上長途跋涉而來、終於落地，抵達目的地的空港或海港之前的所有成本，包括運費、保險、倉儲等，有些還會將關税等税負加上。若是經銷商有經驗，能將landed cost計算得一清二楚，使得「The landed cost was exactly what we had expected.」，讓實際的落地成本就跟原先所預估的一樣，這樣一來，預計的獲利率無須下修，產品利潤可以掌握得更好，一切都照計畫走，心裡上那該有多踏實啊！

Show Time! 換你上場！

❶ 我們是有辦法再降些價格，但也希望您們在獲利率上也能有所調整。

We're able to lower our price to you, but we also expect you to adjust your ____________ ____________.

❷ 我們先前報給客戶的價格已經快跟我們的落地成本差不多了，幾乎是零利潤。

Our previous offer to customer is already close to our ____________ ____________ and we got almost zero margin.

❸ 為了要拿下這個案子，我同意再多給5%的折扣。

____

____

❹ 然而，為表善意，若你們最少能訂到3組，則我們能再多給5%的折扣。還請告知是否要我們提供正式報價單

However, ______ ______ ______, we can offer ______ ______ ______ for a minimum order of 3 kits.

來對對答案

❶ profit margin

❷ landed cost

❸ In order to win this project, I could approve another 5% discount.

❹ in good faith；another（或additional）5% discount

數學時間

【情境】宏大電機公司是製造自動門的臺灣廠商，新研發設計的HT-101 Model旋轉門就快製造出來了，產品經理Janice現正要好好理一理定價，於是，她將所有的成本資料都調了出來，整個加總之後，得出總成本為美金18,000元，那定價要定在什麼水準呢？在她用著她纖纖細指輕敲桌面時，她想到了成本加成定價法／Markup pricing／Cost-plus pricing，不過，加成比例（markup percentage）跟所謂的獲利率（profit margin）好像並不一樣？！怎麼個不一樣呢？二話不說，嗯……一話也不多說，立馬列個表，好好來「克漏格」，先把所有的數值與比例填上再說！

成本：**18,000 USD**		
Markup percentage	**List price**	**Profit margin**
10%		
20%		
30%		
40%		
50%		

如何？填好了嗎？Janice已幫我們算出了正確答案了，我們一起來看看細部數據囉！

Markup percentage	List price / Retail price	Profit	Profit margin
10%	19,800 USD (18,000 × 1.1)	1,800 USD (19,800 - 18,000)	9.09% (1,800 ÷ 19,800)
20%	21,600 USD (18,000 × 1.2)	3,600 USD (21,600 - 18,000)	16.67% (3,600 ÷ 21,600)
30%	23,400 USD (18,000 × 1.3)	5,400 USD (23,400 - 18,000)	23.06% (5,400 ÷ 23,400)
40%	25,200 USD (18,000 × 1.4)	7,200 USD (25,200 - 18,000)	28.57% (7,200 ÷ 25,200)
50%	27,000 USD (18,000 × 1.5)	9,000 USD (27,000 - 18,000)	33.33% (9,000 ÷ 27,000)

在你自己算過之後，相信以後你對加成比例和獲利率的計算就輕鬆自在了！大原則，加成從「成本」加上去，獲利率呢？就得要從「售價」算下來了，所以獲利率永遠比加成比例來得小，因為獲利率是將利潤除以金額較大的定價／零售價呢！

Chapter 1 搞定報價

1-6 下單前所需表格

Tina: Hello, Andy. It's Tina from Leader Biotechnology.

Andy: Hi, Tina. I'm glad you called. Are you going to tell me your order is ready to be placed to us?

Tina: I hope so too, but our customer still needs to get formal approval from his supervisor.

Andy: I see. Does your customer need us to provide any documents?

Tina: Exactly! You're so clever! Our customer needs to have quotation or Proforma Invoice. Also the specification and product manual are requested to be submitted.

Andy: I can forward you a formal quotation on our AB Scientific's letterhead. About the specification, the one I'll send you is the product's datasheet which contains all detailed specifications. As to the mentioned product manual, I assume you refer to the operating manual which includes important details about safety, as well as recommendations for proper use, right?

Tina: Yes! That's the document our customer needs.

Andy: So I'll go prepare these 3 documents and Email them to you later today. Is that okay?

Tina: Perfect! Thank you, Andy!

背景說明	
人物	Tina：經銷商業務代表 Andy：原廠客服代表
對話主題	下單前要先看看產品資料： 客戶下單前需要提送相關文件以供審核。

譯文

蒂娜：嗨，安迪，我是力得生物科技的蒂娜。

安迪：嗨，蒂娜，很高興妳來電。是不是要告訴我你們的訂單已準備好要發給我了呢？

蒂娜：我也希望是這樣，不過，我們的客戶還得要取得主管的正式核可。

安迪：瞭解，所以妳的客戶有要我們提供任何文件嗎？

蒂娜：一點也沒錯！你真的是太聰明了！我們的客戶需要報價單或形式發票，另外也需要規格書跟產品手冊。

安迪：我可以轉一份有我們AB科技公司信紙抬頭的正式報價單給妳。關於規格書，我會寄給妳的資料是產品的説明書，裡頭有所有詳細的規格説明。至於妳説到的產品手冊，我想妳指的是操作手冊，它裡面有關於安全的重要細節，還有對正確使用的建議事項，是這份嗎？

蒂娜：對！那就是我們客戶要的。

安迪：所以我去準備這三種文件，然後今天稍晚就Email給妳，這樣可以嗎？

蒂娜：太好了！謝謝你，安迪！

例一 | 可給形式發票，以確認訂購，並供預付貨款。

例句　Please let me know if you are interested in ordering and, if yes, we will send a Proforma Invoice with bank details for your prepayment.

譯文　請告知是否您有興趣下單，若有，我們將發送形式發票給您，上頭也會有銀行詳細資料，供您預付貨款用。

例二 | 給形式發票，供預先付款。

例句　Thank you for your enquiry. As you have not ordered from us before, we would supply to you on a Proforma Invoice basis. Upon receipt of payment, your order will be dispatched.

譯文　謝謝您發來詢價，因為您未曾跟我們訂過貨，我們會以形式發票的方式來供貨給您，等收到您的貨款之後，我們即可安排出貨。

例三 | 給形式發票，確認進口不需要許可證，確認形式發票有簽回後，就可出貨。

例句　Please find attached our Proforma Invoice in relation to your recent enquiry. Once we have received confirmation of whether an import permit is required and your signed proforma for acceptance, we can dispatch for you.

譯文　對於您最近的詢價，我們在此提供形式發票如附，等收到您回覆確認是否需進口許口，及收到您簽回形式發票，表示接受之後，我們就可安排出貨。

例四 | 給報價單，可填妥回傳，完成訂購。

例句 Thank you very much for contacting us. I've attached the quotation for your review. If you would like to purchase, please Email/fax your purchase order to us or by completing the information on the attached quotation and Email it back to me.

譯文 很謝謝您與我們連絡，在此附上報價單供您參考，若您要購買，請Email／傳真您的訂購單給我們，或是在所附的報價單上將資訊填寫完畢，再以Email發回給我。

例五 | 給報價單，保留了貨，下單時加註單號。

例句 Thank you for your price request. Attached is our quotation to this product. You can make your own offer to your customer. We can hold this material until September 30. Please refer to the quotation number when placing your order. Thank you for your continued support.

譯文 謝謝您發來詢價，在此附上此產品的報價單，您可自行報價給您的客戶。貨物可保留至9月30日，下單時，請加註此報價單號。謝謝您一直以來對我們的支持。

 關鍵字急救站—Proforma Invoice

「Proforma Invoice」有它特別的地方，特別在哪裡呢？做國貿的人都知道它，也常會碰到這份文件，但很多人都不知道要怎麼說這份文件的中文名呢！它從字面上來翻譯，就叫「形式發票」，從功能上來說，也可稱呼它為：預估發票、預開發票、估價發票，是原廠與供應商在訂單前置溝通階段時，提供給買方的一份參考性質的發票，上頭列有產品型號、品名、規格、單價等資訊。因是參考，因還不是正式發票，所以在Invoice前頭加了個「Proforma」，這個字也常寫為Pro forma兩個字，是拉丁文，意思為「for the sake of form」／為了形式，所以Proforma Invoice是一種為了形式、純為形式的發票，並無發票的實質意義，是賣方要銷售產品給買方之前，先開立的一種發票文件，讓買方藉以估計進口成本，所以也稱為估價發票、試算發票，所以，若你查英英字典，會看到Proforma Invoice的解釋為：「a quotation that tells a customer how much a particular piece of work would cost」，功能上也就跟報價單一樣，的確，在實務運作上，有的原廠與供應商也就直接將Proforma Invoice當作報價單來開立，而且還會請買方若是確定要訂購，就請直接在Proforma Invoice上簽名回傳，這樣即可當作確認下單，達成交易，賣方也就會去處理訂單、安排出貨了！

術語直達車，專業補給站

原廠在客戶下單正式訂購之前，會需要出具、提供那些文件呢？價格重要，而產品對不對、合不合客戶所需也重要，所以在客戶詢價時、下訂單前的前置作業階段，原廠會要提供給客戶的文件就包括有：

一、Quotation／報價單 & Proforma Invoice／形式發票

一份完整、清楚的報價單或Proforma Invoice，一定要包括下面各點資訊：

❶ Business details／商業明細

報價單要出具給誰，一定會列在報價單上，所以賣方會請買方提供連絡資訊，或是提供billing address（發票地址）、shipping address（出貨地址）、Email address及phone no.。

❷ Product details／產品明細

這部分的資訊包括有產品的Catalog no.／型號、Product Description／品名，以及Size or Specification／規格。

❸ Price details／價格明細

各項產品的單價、小計，以及要收取的各項相關費用，都得在報價單或Proforma Invoice一一列出，以免客戶到了要下單時，才發現原廠還有一些滴滴落落的雜費要收！

❹ Payment terms and conditions／付款條件

這部分須列出付款上的要求，看是要求客戶預付貨款，或是收貨時再付款等等不同的付款條件，還是要客戶選擇匯款或刷卡的付款方式。

❺ Quote expiry date／報價效期

原廠在出具報價單或Proforma Invoice時，一定要記得加上報價效期，一方面對報價可有較好的管控，不至於在好些時日之後，當報價

狀況、條件已然改變時，還得依當初報價單所列價格來執行訂單。另一方面，列了效期的報價單也可吸引客戶快些下單。

❻ Customer acceptance signature／客戶簽名確認

若要讓客戶能夠簽回，以確認訂單，則請在報價單或Proforma Invoice上列出這樣的句子：「I, [name] , accept the above terms and conditions. Signed ____________ Date ____________」／「我，〔姓名〕，接受上述所列條件。簽名____________ 日期____________」

二、產品資料

客戶在評估採購案時，會要求原廠提供產品的相關資訊，供其參考與評估，以決定產品是否合用。這樣的產品資料可還不少，依產業不同也各有所用，在此我們就列出常見的一些文件來瞧瞧囉！

- Datasheet／產品說明書、產品規格書
- Specification／產品規格書
- CoA／Certificate of Analysis／產品分析報告
- Package insert／仿單（附在產品包裝內的使用說明書）
- Operating manual、User manual／操作手冊、使用手冊

接下來，我們就來看看報價單與Proforma Invoice的範例格式囉！

MyTech Chemicals

QUOTATION

No. 1202, Sec. 5, Hsin-Yi Rd., Taipei, Taiwan
www.mytechchem.com.tw

Phone: 886-2-2799-1188
Fax: 886-2-2799-1190
sales@mytechchem.com.tw

DATE	2023/5/30
QUOTATION #	A0202
Customer ID	7261

PLEASE FILL OUT YOUR BILLING AND SHIPPING ADDRESS PRIOR TO SUBMITTING YOUR ORDER

BILL TO:

Gary Collins
Amber Science, Inc.
9250 Pacific Heights Blvd.,
San Diego, CA 92121, USA
858-455-9250
gcollins@gmail.com

SHIP TO (if different):

[Name]
[Company Name]
[Street Address]
[City, ST ZIP]
[Phone]
[Fax/Email]

Payment Information

Credit Card #:
Name on Card:
Expiration Date:
Card Verification #:

Other Information

PO #

FedEx #

ITEM #	DESCRIPTION	QTY	UNIT PRICE	TOTAL
MTC0925	NR-1 Antibody ELISA Kit	5	865.00	4,325.00
	Bank Name: Taiwan Bank, Cheng Chong Branch			-

	Address: 47, ChinTau E. Rd, Taipei, Taiwan, R.O.C.			-
	Telephone: 886-2-2321-8934			-
	Account Name: MyTech Chemicals			-
	Account Number: 045007019250			-
	Swift Code: BKTWTWTP045			-

SUBTOTAL	$4,325.00
S & H	170.00
TOTAL USD	$4,495.00

Other Comments or Special Instructions
Valid for 30 days from the date of this quote. If using your own FedEx #, please remove S & H fee and add $25 for packaging & handling. Please reference quote # if you wire transfer. *ALL PRODUCTS ARE FOR RESEARCH USES ONLY.*
To accept: (1) Please write your name and sign above
(2) Email to sales@mytechchem.com.tw OR Fax to 886-2-2799-1190

If you have any questions about this quotation, please contact
Ann: Achen@mytechchem.com.tw

Thank You For Choosing MyTech Chemicals!

邁科化學

報價單

臺灣臺北市信義路五段1202號9樓
電話：886-2-2799-1188
傳真：886-2-2799-1190
www.mytechchem.com.tw
請在發送訂單前，填寫您的發票地址與出貨地址。

日期	2023/5/30
報價單號	A0202
客戶編號	7261

發票：

蓋瑞・柯林斯
安柏科技公司
美國加州聖地牙哥 CA 92121
太平洋高地大道9250號
858-455-9250
gcollins@gmail.com

出貨：（如與發票地址不同）

[姓名]
[公司名]
[街道地址]
城市、州、郵遞區號
[電話]
[傳真／Email]

付款訊息：

信用卡卡號：
持卡人姓名：
到期日：
驗證碼：

其他訊息：

PO #

FedEx #

型號	品名	數量	單價	總價
MTC0925	NR-1 抗體酵素免疫分析法試劑組	5	865.00	4,325.00
	銀行名：臺灣銀行城中分行			-
	地址：臺灣臺北市中正區 青島東路47號			-
	電話：886-2-2321-8934			-
	帳戶名：邁科化學			-
	帳號：045007019250			-

	銀行國際代碼：BKTWTWTP045			-

小計	$4,325.00
運費與處理費	170.00
總計 USD	$4,495.00

其他訊息或特別指示
此報價自報價日起30天有效。 若您欲以自己的FedEx帳號出貨，請刪除運費與處理費，改加計包裝費及處理費 25元。 匯款時請加註此報價單單號。 *所有產品僅供研究用。*
回覆確認接受：__________________ (1) 請在此寫上名字並簽名。
(2) 請以Email回覆至：sales@mytechchem.com.tw或傳真至886-2-2799-1190

若您對此報價有任何問題，請連絡

Ann: Achen@mytechchem.com.tw

謝謝您選擇邁科化學！

PROFORMA INVOICE

Hi-Tech Biomedicals	PROFORMA INVOICE #:	0034891
925, Sec. 5, Hsin-Yi Road,	DATE:	06/30/2023
Taipei, Taiwan	CONTACT PERSON:	Hank Chen
Tel: 886-2-2769-0000	CUSTOMER #:	1202
Fax: 886-2-2769-0001	CUSTOMER FAX:	858-455-9255

Email: hankc@hitechbiomed.com.tw
Website: www.hitechbiomed.com.tw

SOLD TO:	SHIP TO:
Gary Collins	Gary Collins
Amber Science, Inc.	Amber Science, Inc.
9250 Pacific Heights Blvd.,	9250 Pacific Heights Blvd.,
San Diego, CA 92121, USA	San Diego, CA 92121, USA
Tel: 858-455-9250	

CUSTOMER P.O.	SHIP VIA	F.O.B.	PAYMENT TERMS
1216003	FedEx 881211734	Taipei	CASH IN ADVANCE

Item No.	Ordered	Shipped	Back ordered	Price	Amount
1128-8 Streptomycin	1	0	0	$445.00	$445.00
1127-8 Penicillin	1	0	0	$445.00	$445.00
Bank fee	1	0	0	$30.000	$30.000

Hi-Tech Biomedicals cannot accept any return or exchange after shipment since the returned kits cannot be resold as it is difficult to guarantee their conditions and performance.

Thank you for your order. Please wire the total amount in US Dollars to the following account and Email fax the documents to us. Please pay your wire fee to our bank.
Bank name: Taiwan Bank, Cheng Chong Branch
47, ChinTau E. Rd, Taipei, Taiwan, R.O.C.,
Account owner: Hi-Tech Biomedicals
Account Number: 115092100
Swift Code: BKTWTWTP045

Net Order:	$920.00
Less Discount:	0.00
Freight:	0.00
Sales Tax:	0.00
Order Total:	$920.00

Approved By:________________ Date: ________________

Authorized Signature

Customer's Name
& Signature: ________________

IMPORTANT: Please check all the above details one by one carefully then sign and Email/fax back to us. After confirmation, the order can not be cancelled.

形式發票

高德生醫

形式發票號碼：0034891

臺灣臺北市信義路五段925號

日期：06/30/2023

電話：886-2-2769-0000

連絡人：Hank Chen

傳真：886-2-2769-0001

客戶編號：1202

Email：hankc@hitechbiomed.com.tw
網址：www.hitechbiomed.com.tw

客戶傳真：858-455-9255

銷售至:

蓋瑞．柯林斯

安柏科技公司

美國加州聖地牙哥 CA 92121

太平洋高地大道9250號

電話: 858-455-9250

出貨至:

蓋瑞．柯林斯

安柏科技公司

美國加州聖地牙哥 CA 92121

太平洋高地大道9250號

客戶訂單號碼	出貨經	F.O.B.	付款條件
1216003	FedEx 881211734	臺北	出貨前現金付款

型號	已訂	已出	缺貨待出	單價	總價
1128-8 鏈黴素	1	0	0	$445.00	$445.00
1127-8 青黴素	1	0	0	$445.00	$445.00
銀行手續費	1	0	0	$30.000	$30.000

產品一經出貨，高德生醫即不接受任何退貨或換貨，因退回產品之狀況與效能難以保證，故無法再銷售。

謝謝您的訂單，請匯美金總額至以下所列帳戶，並Email／傳真文件給我們。匯款手續費請您支付給銀行。
銀行名：臺灣銀行城中分行
臺灣臺北市青島東路47號
帳戶戶名：高德生醫
帳號：045007019260
銀行國際代碼：BKTWTWTP045

訂單淨額：	$920.00
折扣：	0.00
運費：	0.00
銷售稅：	0.00
訂單總額：	$920.00

核准：＿＿＿＿＿＿＿＿ 日期：＿＿＿＿＿＿＿＿
授權人簽名

客戶名與簽名：＿＿＿＿＿＿＿＿

重要：請逐一仔細檢查上述明細，簽名後以Email或傳真回傳給我們。訂單一經確認後即無法取消。

Show Time! 換你上場！

❶ 請告知您是否有興趣下單，若有，我們將寄形式發票給您，上頭也會有銀行的詳細資料，供您預付貨款之用。

Please let me know if you are interested in ordering and, if yes, we will send a ________ ________ with bank details for your ________.

❷ 我們的客戶需要報價單或形式發票，另外也要求規格書。

Our customer needs to have ________ or ________ ________. Also the ________ is requested to be submitted.

❸ 等我們一收到您簽回形式發票表示接受之後，我們就可安排出貨。

____

____

❹ 謝謝您對我們的產品有興趣，請見我們的形式發票與產品說明書如附。

____

____

____

來對對答案

❶ Proforma Invoice；prepayment

❷ quotation；Proforma Invoice；Specification

❸ Once we have received your signed Proforma Invoice for acceptance, we can dispatch for you.

❹ Thank you for your interest in our products. Please find attached our Proforma Invoice and data sheet.

Chapter 2 篇章簡述

當買賣雙方終於確認了所需的產品與規格，也談妥了價格之後，就來到下單這一步了。在下單前後，分別需要注意哪些事情呢？哪些資訊與文件一定要齊備呢？我們在這一節當中，就來詳細話說分明囉！

賣方若已報了價，但買方遲遲沒有進一步的消息，賣方這時就得「適度」地主動出擊，問一問報價結果，探一探案子狀況，這樣才能在還有辦法施力的時間點上多談一些，增加成交的可能性。等到買方真正答應要下單了，那接下來的事就簡單多了，最主要的就是請客戶將如下的訂單明細說清楚、寫齊全：

➡ 公司資料

- Name of institution or company／機構或公司名
- Account number (if available)／買方帳號（如有申請）
- Shipping address／出貨地址
- Billing address／發票地址
- Name and phone number of distributor／經銷商的連絡人姓名與電話
- Name and phone number of the end user／最終使用者的姓名與電話

➡ 產品資料

- Quotation number／報價單單號
- Catalog number and description／型號與品名
- Quantity and size required／所需數量與規格
- Price／價格

➡ Terms of payment／付款條件

待賣方需要的這些下單資訊都到位了之後，接著就是要賣方出具「訂貨確認單」給買方了。在「訂貨確認單」中，上述各個要項一樣會出現，另外，在付款條件上，賣方通常也會再加上付款方式的選項。

在這一章中，我們就一起從訂單前的催促下單開始，告訴你可以怎麼催，有哪些英文說法，接著就來說明「錢」事，瞧瞧付款條件有哪些，付款方式有哪幾種，最後再來實地看看訂貨確認單的內容與格式，讓我們將訂單的相關知識穩穩地「釘」下來！

Chapter 2 搞定訂單

2-1 催促下單

對話 MP3 07

Peter: Anderson Group. Good morning. This is Peter Tyler. How may I help you?

Julie: Hi Peter! This is Julie Chen from Sunrise Enterprise. I'm calling to follow up on the bulk order that we discussed last week.

Peter: That's the thing we're working on right now! I just had a question. What are the terms of payment?

Julie: For international orders, we always ask our customers to make advance payment.

Peter: I see. Do you accept credit card?

Julie: Yes, we do. If you prefer, you could pay by wire transfer as well.

Peter: Thanks for letting me know about this. We'll compare their handling fees and then decide the way of our payment.

Julie: No problem. By the way, are there any special documents or procedures required for importing into your country?

Peter: No, the ordered product is not regulated in our country. We don't need to obtain an import permit for it.

Julie: Good. Please send me your order no later than this

	Friday. Our campaign price will be expired after that day.
Peter:	**Okay, I've noted this deadline.**
Julie:	So I'll let you go to work on the order for us!
Peter:	**Hahhah...That's very thoughtful! Bye! Have a nice day!**

背景說明	
人物	Peter：經銷商業務人員 Julie：原廠業務人員
主題	要下單了嗎？要下請快！ 買方確認付款條件與方式，賣方提醒報價效期。

譯文

彼得： 安德森集團，早安，我是彼得・泰勒，有什麼我能替您服務的嗎？

茱莉： 嗨，我是日昇企業的茱莉，我打來是要來問問我們上星期談到的那份大訂單的狀況。

彼得： 我們正在辦這件事呢！我剛好有個問題，請問付款條件是什麼呢？

茱莉： 對於所有的海外訂單，我們都是要求客戶要預先付款。

彼得： 瞭解，那妳們接受信用卡嗎？

茱莉： 可以的，若你們想要用匯款支付，我們也可接受。

彼得： 謝謝妳告訴我這個訊息，我們會比較一下兩個付款方式的手續費，之後再來決定我們的付款方式。

茱莉： 還有一件事，請問你們進口這項產品需要申請什麼特殊的文件，或有什麼特別的程序要處理嗎？

彼得： 沒有，我們訂的這項產品在我們國家並沒有管制規定，不需要申請進口許可證。

茱莉：好的，那就請你在星期五前發訂單給我了，我們的促銷特價過了星期五就失效了喔。

彼得：好，我會記著這個截止日期。

茱莉：那麼我就讓你去忙我們訂單的事囉！

彼得：哈哈……您設想得很周到呢！再見，祝妳有個美好的一天！

說三道四，換句話試試

例一 | 詢問報價進展。

例句 I am writing to follow up with you on the pending quote. We would be glad to help you in placing an order.

譯文 我想問問先前給您的報價有沒有進一步的消息，我們很樂意來協助您下單。

例二 | 詢問一下報價結果。

例句 I am just following up on the quote #Q409563. If there is anything we can do to be of further assistance, please feel free to contact us.

譯文 我想來跟進一下報價單單號Q409563的結果，若是有我們可以進一步協助的地方，請儘管與我們連絡。

例三 | 有收到報價嗎？有問題就請提問。

例句 I am writing to check whether you have received our previous quote as attached. I was wondering how it is going on now. If you have any questions or need any help, please do not hesitate to contact me at 886-2-8559-9250 or by Email. Have a nice day! Looking forward to your order!

譯文 請問您有沒有收到我們先前所發來的報價（如附），我想問問看目前進展如何，若是您有任何的問題或是需要什麼協助，請儘管打個電話給我，號碼為886-2-8559-9250，或是用Email跟我連絡也可以。祝您有個美好的一天，期待收到您的訂單！

例四｜您的客戶決定下單了嗎？

例句 I am following up on our recent quotation for E-101 Model. Is your customer interested in placing the order? Please let me know the status of the quote so I can update my file. Thank you.

譯文 我想要來問問我們最近所報機型E-101的報價結果，您的客戶有要下單了嗎？請告訴我報價的狀況，讓我能夠更新一下我的檔案紀錄，謝謝。

例五｜這案子還有沒有機會？

例句 I hope you are well! I wanted to follow up on your E-101 Model inquiry last month. Does this opportunity still exist? Do you have any questions that I can help answer? Let me know what I can do to help you.

譯文 收信好！我想來問問您上個月對機型 E-101所提詢價的進展，這案子還有什麼機會嗎？有什麼問題需要我幫忙回答的嗎？若有我能協助的地方，就請告訴我。

關鍵字急救站—bulk order

「Bulk order」是個原廠或供應商一看到就會開心的詞兒！有order好，有bulk order這種大訂單，那就更是好得不得了！經銷商或客戶當然也喜歡有bulk order，因為訂單大，就有更多的籌碼來好好協商與談價一番，以爭取到更好的條件！而bulk這個字，在協商訂單、要求折扣時，可是這兒bulk、那兒bulk，說法還頗多呢！所以，我們就來好好報報bulk囉！

bulk [bʌlk] adj.
relating to the sale, production, or transport of goods in large quantities／商品在產銷運輸上大量、大批、大宗的

我們要吸引客戶多訂些貨，就會制定出bulk discount／量大折扣，希望客戶爽快地來做筆bulk buying、bulk purchase／量大採購，直接下個bulk order／大訂單！若是客戶所詢的是小包裝的產品，我們就可以好康逗相報，跟客戶這麼說：

➡ This product is available in 100 μg quantities, with bulk discounts available.
這個產品有100微克規格的量可供應，並且提供量大的折扣。

或者也可以告訴客戶：

➡ If you are interested in bulk pricing or our custom services please contact us.
若是您對量大價格或我們的訂製服務有興趣，就請跟我們連絡。

此外，在下訂單時，賣方會希望客戶若有大單要下，最好早點告知，讓賣

方及早準備，因此，就會有這類的要求：

➡ If you have bulk orders, please submit the majority of your weekly orders on Thursdays by 7:00 am with any last minute, smaller requests submitted on Fridays. This will help to guarantee shipments of your bulk orders and alert you earlier of any potential backorders.

若您有大訂單，請在星期四7:00 am 前將大部分的訂單先發來給我們，星期五可發給我們最後要追加、較小量的訂單。這有助於保證您的大訂單可準時出貨，若有任何可能的缺貨品項，我們也可及早告訴您。

這樣的接單規定是一定得要清清楚楚地告訴客戶，讓客戶知道再怎麼忙，也要在某個時間點之前下單，容不得左拖右拉，若是晚了接單時間，就是損了自己的權益。賣方在bulk order的處理要求上，一定要有明白的規定，給予自己充裕的時間，以能在bulk order這種量多、價高案件的處理上，能有高規格的服務品質表現！

術語直達車，專業補給站

「催客戶下單」，雖說是「催」，其實是請求，請求客戶早些捧著訂單、送給賣方，所以這樣子的「催」，可就不是誰落了進度、說了沒做、做了做不完整這類的問題，也不是歸責後的應對處理。既然說催人訂單不是那種催，在做法上那種催跟這種催在催法上會有什麼不同嗎？有滴！請求下單的催法，「情理」要顧，顧完之後再顧「事理」！所謂的「情理」，說的就是關於催促下單的基本禮儀與態度，我們就一起來檢視自己有沒有做到這樣的「情理」囉！

情理一：有什麼需要協助的嗎？

在賣方報價之後，客戶卻遲未回覆、未下單，表示客戶心中有些想法，可能有競爭者給了更好的條件，所以客戶並未將你列為協商對象的第一個順位，這時呢！請及早也適時跟客戶連繫，問問客戶的「feedback」與「interest」，看看價格上有無需要協商的地方，或是有其他條件需要再討論。

情理二：什麼時候催？多久催一次？

在這個資訊快速流通的世代，企業要贏得客戶，必得快速、靈活地因應客戶所需，所以當客戶遲未下單，我們一定要天天催……喔不！若你天天催，只會催出客戶的反感啊！請兩、三天一小催，催個兩次之後，若還沒啥確切的訊息，那可能客戶下單的機率就小了，此時，蒐集資訊的功能反而強過接單實績了。接著，我們就來看看幾種催促的問法：

➡ I was wondering if you are going to place the order for these products. Please let me know any feedback related to this quotation.
您是否有要下單訂購這些產品呢？還請告知關於這個報價的任何回饋訊息。

➡ Have you got any feedback from your customer? Is there anything else we can do for you? We look forward to your early reply.
請問您的客戶有回饋什麼樣的消息嗎？有什麼我們可以為您服務的嗎？期待盡快收到您的回音。

➡ Have you had a chance to review the price we quoted? Can we be of any further assistance? If yes, please do not hesitate to contact us.
您是否已有機會評估一下我們所報的價格了嗎？有什麼我們可以協助的地方嗎？若有，還請儘管與我們聯絡。

說完情理後，那催促下單的「事理」是指什麼呢？

事理一：報價效期快到了！

報價單上多會列明效期，因此，在催促客戶下單時，則可告訴客戶：

➡ Please be reminded that our quote will be expired next week and so please place your order before the deadline.
在此提醒一下，我們報價的效期只到下星期，所以還請您在到截止日之前儘早下單。

事理二：有特殊情況！

當發生客戶所詢產品也有其他的客戶表示興趣時，這就給了賣方一個很好催、有理催的事理了。例如另有客戶詢大單，在存量有限的情況下，確實得要好好來催一催客戶，請他們要下單請早，以免向隅，此時就可以這麼說：

➡ We now have another customer interested in a large number of the same product. Please let us know as soon

as possible whether your customer wants this particular lot or if they could wait for a new one? For your information, the new lot won't be released until the end of next month.
我們現在有另一個客戶對同樣的產品有大量的需求，還請盡快告訴我們是否您的客戶有要這個特定批次的貨，或是他們可以等新批呢？跟您說一聲，新批要到下個月月底才可供貨。

就算我們個性沉穩，該催時還是得加足馬力催下去，但請記得要「好好催」，情理要好，事理要足，才能催來得體，催來有力，催出好結果喔！

Show Time! 換你上場！

❶ 我打電話來是要來問問我們上星期談到的那份大訂單的狀況。
I'm calling to follow up on the __________ __________ that we discussed last week.

❷ 請你在這個星期五前發訂單給我，我們的促銷活動特價過了星期五就失效了喔。
Please send me your order ______________________________.
Our campaign price will be expired after that day.

❸ 請告知報價的狀況，好讓我能夠更新一下我的紀錄。
Please let me know ______________________ so I can update my file.

❹ 我只是想問問看報價單單號Q409563的結果，若是有什麼我們可以進一步協助的地方，請儘管與我們連絡。
____
____

來對對答案

❶ bulk order

❷ no later than this Friday / by this Friday / before this Friday

❸ the status of the quote

❹ I am just following up on the quote #Q409563. If there is anything we can do to be of further assistance, please feel free to contact us.

搞定訂單

2-2 付款條件

Roger: Hello, Jenny. This is Roger from Cosmos Biotech.

Jenny: Hi, Roger. Nice to hear from you. How've you been?

Roger: Not bad. I'm calling to check with you whether you've received our wired payment.

Jenny: Yeah, we did! Our bank just notified us that we had received the money into our account. Thanks.

Roger: That's good. Taking this opportunity, I'd like to discuss with you about payment terms.

Jenny: I'm listening.

Roger: We'd like to know if it's possible to change our payment terms to Net 30 days, instead of advance payment that we have now.

Jenny: Our company policy is to have the prepayment confirmed prior to shipping. However, we do appreciate your hard work in promoting our products. So, for your company, we'll honor your proposed payment terms.

Roger: Great! We'll show the new payment terms on our next Purchase Order.

Jenny: Okay! I'll also send a notification to our accounting department about this change.

Roger: Thanks!

背景說明	
人物	Roger：經銷商業務人員 Jenny：原廠業務人員
主題	討論更改付款條件： 預付貨款 → 淨30天

譯文

羅杰： 珍妮，妳好，我是宇宙生技的羅杰。

珍妮： 嗨，羅杰，很高興接到你的電話，你好嗎？

羅杰： 還不錯，我打來是要問問妳有收到我們電匯的貨款嗎？

珍妮： 有，收到了，我們的銀行剛通知我們，說匯款已經入了我們的帳戶。

羅杰： 那就好。藉這個機會，我想要跟妳討論一下付款條件的事。

珍妮： 請說。

羅杰： 我們想知道有沒有可能將付款條件從目前的預付貨款改成淨30天呢？

珍妮： 我們公司的政策是要出貨前完成預付，不過，我們很感謝你們在我公司產品推廣上所做的努力，所以，對你們公司，我們可依你所提議的付款條件來辦理。

羅杰： 太棒了。我們在下次的訂購單上就會寫出這個新的付款條件囉。

珍妮： 好的。我也會發通知給會計部，讓他們知道有這項變更。

羅杰： 謝謝了！

說三道四，換句話試試

例一｜就是要預付貨款，其餘免談！

例句一 For all international orders, our company requires prepayment always.

譯文一 對於所有的海外訂單，我們公司都是要求預付貨款。

例句二 We accept only full payment up front for international orders.

譯文二 對於所有的海外訂單，我們只接受預付貨款。

例二｜新客戶得要預付貨款後，原廠才會生產新批。

例句 As you are our new customer, we will require your payment up front before we could produce the new batch.

譯文 因為您是我們的新客戶，我們會要求您在新批生產前，預先支付貨款。

例三｜預付貨款後，就會安排出貨。

例句 Prepayment is required on all international shipments. Once our bank confirms that your wire transfer is received, shipment will be made on the next possible Friday.

譯文 所有的海外訂單皆須預付貨款，等我們的銀行確認收到電匯匯款後，我們就會在下一個星期五安排出貨。

例四｜要信用狀才有信用！

例句　We prefer letters of credit for our international orders.

譯文　我們對海外訂單多是採用信用狀的付款條件。

例五｜改付款條件

例句　For international orders, we always ask our customers to make the entire payment in advance. But, as a gesture of goodwill, I could make an exception to you and accept "NET 30 days" this alternative way of payment.

譯文　對於海外訂單，我們都會要求客戶預付貨款，不過，為了表示善意，我可破個例，接受「淨30天」的這個不同的付款條件。

例六｜要貨到前付款，不要貨到後付款。

例句　All orders must be prepaid. We are unfortunately unable to accept cash on delivery.

譯文　所有訂單皆須先付款，很可惜我們無法接受貨到付款。

關鍵字急救站—payment

說到與付款有關的事，一定會用到「payment」這個字。這個字簡單啊！就是動詞「pay」加上名詞字尾–ment就成啦！是的，就是這類這麼簡單的字，變化起來才輕盈自在呢！我們現在就來自在地看看「pay」可以變出什麼樣的專業內涵囉！

Payment／付款、貨款

說到貨款或付款的金額與動作，就是payment 現身的時候了！若要請客戶在交貨後，以電匯支付全額貨款，就要這樣說：「100% payment needs to be made on delivery by wire transfer.」，若要跟客戶說因還沒收到貨款，所以還沒出貨，就可說：「We haven't shipped indeed but it's because we haven't receive your payment.」。要說付款條件，就是payment terms或terms of payment了，若要催客戶付款，提醒客戶履行30天的付款條件，請他們儘速支付貨款，就要這麼說了：「To ensure that you can fulfill the 30-day payment terms, please pay for the following Invoices as soon as possible.」。

Prepayment／預付貨款

要求先看到錢才放貨的付款條件就是prepayment了，這種預付貨款的條件還可以這麼說：Payment in Advance、Payment up front，或是現金預付Cash in Advance〔CIA–你瞧瞧，國貿裡也有中央情報局CIA（Central Intelligence Agency）呢！〕

Prepaid／預付

在填寫出貨提單時，都會看到個幾次「prepaid」這個字，與其搭配的另一個選項則是「collect」／到付。下表即是出貨提單上的部分內容：

Prepaid 預付	Weight Charge 計費重量	Collect 到付	Other Charges Prepaid 其他費用 預付

在國貿上還會常見到「prepaid」的這個搭配詞：「Freight prepaid」，若是買方要求「Freight prepaid」，就是要賣方支付運費，或是賣方支付後，將運費金額加入Invoice中，再跟買方收取。若是買方要求出貨要找其配合的快遞公司，讓快遞公司從買方帳上扣除運費，那就是「freight collect」，賣方無須負擔運費，但需要跟買方索取其快遞帳號，以列在提單上。

Down payment／頭期款

你有沒有想過為什麼頭期款要叫做「Down payment」？「Down」這個字有「立即付現」的意思，「used for saying that you pay an amount of money immediately when you buy something and will pay the rest later」，若我們說買一台新車只要現付多少，之後月付多少，則可這麼說：「You could own a brand new car for only $100 cash down and $1,000 a month.」

Payable／應付的

我們在看Invoices或在催款時，常會用到payable的這個詞組：「Accounts payable」，這是一項會計科目，意思為「應付帳款」，那請你來猜猜與它相對的「應收帳款」的英文要怎麼說呢？ 是的，就是「Accounts receivable」。「Accounts payable」也常出現在訂貨確認單發票上的「Billing information」或 「Bill to」／「發票開至」一欄中，意思即是發票開立的對象為買方的應付帳款部門。下表即是發票上的部分內容：

BILL TO 發票開至
Phoenix Chemicals, Inc. Accounts Payable 1555 Beach Road, Burlingame, CA 94010 鳳凰化學公司 應付帳款部門 美國加州柏林格姆海灘路1555號

術語直達車，專業補給站

在國貿操作實務中，雖然花在付款條件討論上的時間通常沒有比產品詢問、下單、催出貨來得多，但錢這事可是一板一眼的，一開始就要把條件談清，把相關的狀況問個明白，才不會因為付款一事而延誤了出貨或其他時程上的安排。現在，就讓我們來一一了解一下，看看各種付款條件究竟是如何讓貨款從買方的荷包，「錢」進到賣方那兒去！

信用狀／Letter of Credit (L/C)

買方向銀行申請開立信用狀，要將這張附有條件的付款保證文件，開給賣方。銀行向賣方承諾，若賣方能履行所規定的條件，並提示出貨文件，則可擁有開狀銀行的付款擔保。

付款交單／Document against Payment (D/P)

賣方在貨物裝運後，開出匯票，連同出貨文件，委託銀行交給進口地的代收銀行，代為向買方收取貨款，買方須付清貨款後，才能取得出貨文件，辦理提貨清關。

承兌交單／Document against Acceptance (D/A)

承兌交單與上述付款交單的程序相同，唯一不同之處在於代收銀行僅需買方在匯票上承兌，即可取得出貨文件，辦理提貨清關，等到規定的付款期限到時，再行付款。

憑單據付款／Cash against Documents (CAD)

賣方在貨物裝運後，將出貨文件於出口地交給買方或其代理人，即可自買方或代理人處收取貨款。

貨到付款／Cash on Delivery (COD)

賣方在貨物裝運後，將出貨文件交給買方辦理提貨清關後，則可向買方收取貨款。

預付貨款／Cash in Advance (CIA)、Payment in Advance、Prepayment

買方預先支付貨款給賣方之後，賣方即可辦理出貨。

在信用狀、付款交單、承兑交單這幾種付款條件中，銀行這一個中介角色皆有收取、轉交出貨文件或／及匯票的任務。而對其他的付款條件來說，銀行就無此責任了，有的只是處理買方電匯或開支票的要求了。

最後，讓我們來一次盡覽所有常見的付款條件，看看它們的英文名稱、簡稱，以及簡單又扼要的定義囉！

Payment terms 付款條件	Description 說明
Net monthly account 月結	Payment due on last day of the month following the one in which the invoice is dated 發票日隔月最後一天付款
PIA	Payment in advance 預付貨款
Net 7	Payment seven days after invoice date 發票日後7天付款
Net 10	Payment ten days after invoice date 發票日後10天付款
Net 30	Payment 30 days after invoice date 發票日後30天付款
Net 60	Payment 60 days after invoice date 發票日後60天付款
Net 90	Payment 90 days after invoice date 發票日後90天付款
EOM	End of month 月底
21 MFI	21st of the month following invoice date 發票日隔月的21日
1% 10 Net 30	1% discount if payment received within ten days otherwise payment 30 days after invoice date 發票日後10天內付款可享1%折扣，否則為發票日後30天付款
COD	Cash on delivery 貨到付款
Cash account 現金帳	Account conducted on a cash basis, no credit 現金基礎，不接受信貸

Letter of credit 信用狀	A documentary credit confirmed by a bank, often used for export 銀行保兌的跟單信用狀，通常用於出口業務
Bill of exchange 匯票	A promise to pay at a later date 供日後再行付款的承諾
CND	Cash next delivery　下次交貨付現
CBS	Cash before shipment　出貨前付現
CIA	Cash in advance　預付貨款
CWO	Cash with order　訂貨付現
Contra 抵銷	Payment from the customer offset against the value of supplies purchased from the customer 客戶的貨款與供應商自客戶處購買貨品的金額相抵
Stage payment 階段付款	Payment of agreed amounts at stage 依同意之分階段金額來付款

Show Time! 換你上場！

付款條件的頭字語縮寫可真不少，你都記住了嗎？我們來確定一下是否這些字謎我們都解得開！

付款條件簡稱	英文全稱	中文
L/C		
D/P		
D/A		
COD		
CBS		
CIA		
PIA		

來對對答案

如果你寫得順，九成九你會是對的！如果你寫來卡卡，表示你就要獲得長足的進步了呢！請翻翻前面幾頁，找一下答案，讓你自己在「喔～喔～」的瞭解／了然聲中，輕鬆搞定付款條件的名稱囉！

Chapter 2 搞定訂單

2-3 付款方式

對話 MP3 09

Celine: Hello. Is this Peter?

Peter: It is. Who may I ask is calling?

Celine: My name is Celine Chin. I'm in charge of accounts payable here at Energy Healthcare Group. We just ordered A1 Sonic instrument from you.

Peter: Yeah, I'm aware of that order. How may I help you, Celine?

Celine: We noted from your Proforma Invoice that there's a bank fee included, right?

Peter: Yes, for all international orders, there will be the standard 50 USD fee to cover bank charges.

Celine: This amount is high... Do you accept credit card payment?

Peter: Yes, payment can be carried out by bank transfer or credit card through PayPal.

Celine: We do have a PayPal account. We prefer to pay by credit card considering the high bank fee for wire transfer.

Peter: Got it! I'll send you a PayPal invoice and dispatch your order on receipt of payment.

Celine: That will be fine! Thank you, Peter!

背景說明	
人物	Celine：經銷商會計主任 Peter：原廠客戶服務人員
主題	討論付款方式： 匯款 → 信用卡付款

譯文

席琳： 你好，請問是彼得嗎？

彼得： 是的，請問您是哪位？

席琳： 我是覃席琳，在活力保健集團負責應付帳款的業務，我們剛跟您訂購了A1音波儀器。

彼得： 有的，我知道這個訂單，有什麼我可協助的地方嗎？

席琳： 我們看Proforma Invoice上列有一筆銀行手續費的金額，是嗎？

彼得： 沒錯，對所有的海外訂單，我們都會加收美金50元的銀行手續費。

席琳： 這金額高耶…您們接受信用卡付款嗎？

彼得： 有的，我們接受匯款支付，也接受透過PayPal的信用卡付款。

席琳： 我們有PayPal帳戶。因為匯款的銀行手續費高，我們想要以信用卡來付款。

彼得： 收到！我會發一封PayPal Invoice給妳，等我們收到貨款後就會出貨。

席琳： 好的，謝謝你，彼得。

例一 | 可接受的付款方式

例句 Payments can be made by bank transfer, credit card or mailed check.

譯文 可透過銀行轉帳、信用卡或郵寄支票來付款。

例二 | 通知電匯銀行資料

例句一 Here is the information you need to make wire transfer.

譯文一 您匯款會需要的資料在此。

例句二 The wire transfer information is as follows.

譯文二 匯款資料如下所述。

例三 | 可接受信用卡付款

例句一 We do accept credit card payment. Please follow the link below to make payment by credit card.

譯文一 我們確實可接受信用卡付款，請進入下列連結，以進行信用卡付款的動作。

例句二 We can accept credit card payments, all cards except American Express. To pay by card, please Email your card details to receivables@vision.com.tw.

譯文二 我們可接受信用卡付款，除了美國運通卡之外的所有信用卡都可受理。若您要以信用卡支付，請將卡片明細以Email傳至receivables@vision.com.tw。

例句三 For the payment, we can also take credit cards, VISA and MasterCard are both accepted. For more information about payment methods, please look at our homepage www.swan.com.tw.

譯文三 關於付款，我們也可接受信用卡付款，VISA和Mastercard 皆可，若您需要更多關於付款方式的資訊，請參閱我們網頁上的說明：www.swan.com.tw。

例四 | 匯款支付，小額也可接受信用卡付款

例句 We take VISA and MasterCard payment for orders under 1,000.00 USD. You can also pay by wire transfer which related information is listed below.

譯文 訂單金額若小於美金1,000元，則我們可接受VISA及MasterCard信用卡付款，您也可以選擇匯款支付，相關資訊如下所示。

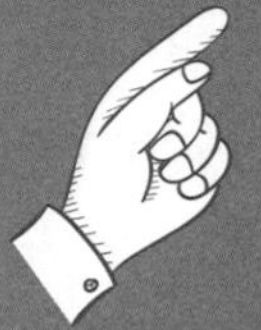

Payment Information
for International Customers

Prepayment Instructions

1. Prepayment is required on all international shipments.
2. Once our bank confirms that your payment has been received, shipment will be made on the next possible Friday. If the product is out of stock, we will inform you the expected shipping date.

Acceptable Payment Methods

We offer three options for payment:

1. Pay by credit card:

You can contact us with credit card information:
Hank Chen: hank.chen@cpy-biolab.com.tw
Tel: 886-2-2243-0000
Fax: 886-2-2243-0001

2. Send check:

Please make the check payable to:
CPY Biolabs, Inc.
Please mail the check to:
3F, No. 555, Pin-Yuan Rd., SanChung Dist., New Taipei, Taiwan

3. Pay by wire transfer

Bank: Bank of Taiwan
Beneficiary: CPY Biolabs, Inc.
Account No.: 045 007 001202
Swift Code: BKTWTWTP045

- Reference: In the "Reference Field" please include our invoice or proforma invoice number for your order.
- All wire transfer charges should be paid by Sender. Any money shortages in the amount deposited into our account compared to our invoice total will be invoiced separately.

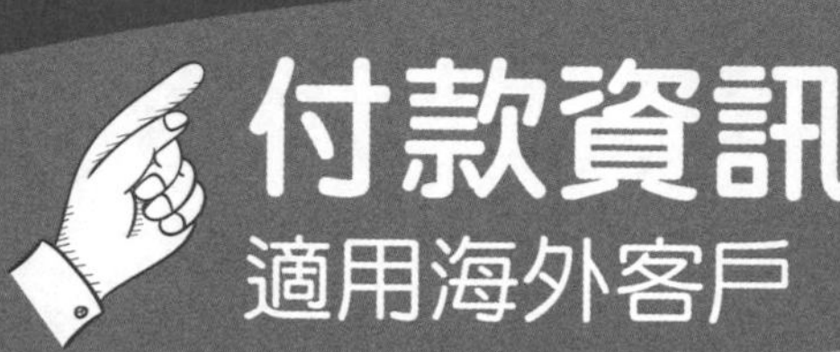

付款資訊
適用海外客戶

預付指示

1. 所有海外出貨皆須預先付款。
2. 當我們的銀行確認收到您的匯款後，我們就會在接下來的星期五安排出貨，若產品沒現貨，我們會再通知您預計的出貨日期。

可接受的付款方式

我們提供三種付款方式：

1. 信用卡支付：

您可與我們連絡，提供您的信用卡資料：

Hank Chen: hank.chen@cpy-biolab.com.tw

Tel: 886-2-2243-0000

Fax: 886-2-2243-0001

2. 支票支付：

支票抬頭請寫：

CPY 生物科技公司

支票請寄：

臺灣新北市三重區平遠路555號

3. 匯款支付：

銀行：臺灣銀行

受益人：CPY 生物科技公司

帳號：045 007 001202

銀行國際代碼：BKTWTWTP045

- 參考：在「參考欄」中，請列出您訂單所對應的我方發票號碼或形式發票號碼。
- 所有的電匯費用應由匯款人支付，若存入我方帳戶的金額少於發票總額，則將另行開立發票收取。

關鍵字急救站

匯款是常見的付款方式，而說到匯款，一定會出現「wire」、「transfer」、「remit」這幾個動詞，其中的「transfer」，我們在Chapter 1的1-1有說明了它「轉帳、匯款」的字義，在這裡我們就一併來看看它與其他兩個字用在匯款時的解釋囉：

- wire (v.): to send money directly from one bank to another using an electronic system（使用電子系統將錢直接從一家銀行寄給另一家銀行）
- transfer (v.): to move money from one account or bank to another（將錢從一個帳戶或一家銀行轉到另一個帳戶）
- remit (v.): to send money to someone, for example as payment for goods or services（寄錢給某人，例如支付貨品或服務的費用）

看完了都是在寄錢、轉錢的這幾個字之後，你有沒有辦法說出它們確實、精確、簡潔的中文字義呢？我們來試試……

- wire有說到電子系統 →「電匯」
- transfer說到move →「轉帳」
- remit就是說send →說寄、發、送都還不夠跟錢有關，所以就是「匯款」囉！

你可能會覺得：「啊電匯、轉帳、匯款不都一樣？」為了驗證是否一樣，我們就得要往深處再探，看看在金融專業層面上，這幾個字的定義是否都一樣，還是有什麼差異囉。

我們先來看看remit的名詞「remittance」，它的定義是這樣的：「It is the process of transferring money from a certain individual to another who is located at a far away place which makes it impossible to give the money in person.」，說的是將錢轉給在遠方的伊人，因距離的因素，例如國外，所以無法當面轉交。要進行remittance，就得借助一個或數個財務機構，有好幾種方法可以轉錢，最常用的方式就是「Electronic Fund Transfer 」（EFT，電子金融轉帳），使用電子資料交換作業，進行資金的轉移及調撥，包括信用卡（Credit Cards）、金融卡（Debit Cards）、企業員工薪資轉帳，以及「wire transfers」。「wire transfers」通常用來指稱一個人轉錢給另一個人的電子轉帳，但嚴格說來，其指的是移轉資金的一種方法，為電子轉帳，通常是從銀行或信用合作社的一個帳戶轉至另一個帳戶。此外，我們在談付款方式時也常會談到「Telegraph Transfer，T/T」，指的也就是國際電匯。

美國聯邦法中有說到「remittance transfers」，這就是囊括美國消費者大多數的電子轉帳，將錢從美國轉給國外的收款人，一般人也稱為「remittances」、「international wires」、「international money transfers」。

對國際貿易來說，收款方都是在國外，在操作上都是透過銀行來轉錢、匯錢，所以wire、transfer、remit這幾個動詞確實可替換使用。最後，我們就來看看這幾個關鍵字的例句吧！

例句一 Thank you for your order. Please wire/transfer/remit the total amount in US Dollars to the following account and

Email the documents to us.

譯文一 謝謝您的訂單，請匯美金總額至下列帳戶，並請Email文件給我們。

例句二 Your invoice is attached. Please wire/transfer/remit payment at your earliest convenience.

譯文二 在此附上您的發票，請盡早匯款。

例句三 Please wire/transfer/remit your payment to Bank of Taiwan...

譯文三 請匯款至臺灣銀行……

術語直達車，專業補給站

在這裡我們接著來說說電匯的實務操作囉！前面說到的「Telegraph Transfer，T/T，國際電匯」，telegraph就是電報，T/T指的就是以電報通訊來完成匯款所需的訊息交換。各國的銀行為了有一個通用的電報訊息規範與訊息交換平台，在1973年成立了全球銀行財務電信協會（Society for Worldwide Interbank Financial Telecommunication, SWIFT），由全球金融業共同合作經營，讓全球的金融機構可以有安全且標準化的信息傳遞服務（例如國際通匯）與介面軟體。進行T/T時，匯款銀行會發出一封SWIFT 格式的電報給匯款銀行的存匯行，再轉往解款銀行的存匯行，或是直接發電報給解款行，以完成匯款程序。以往電匯都是要發電報，但現在大部分地區都以電腦進行銀行之間的轉帳，所以就不再需要電報這個媒介了。

電匯作業完成後，常會碰到的一個問題就是：短收！客戶說明明就已匯了Invoice金額，但賣方收到的金額卻有短少，原因就在於出口業務的國際匯款會由這三方經手處理：匯款行、中間轉匯行、解款行，而這三方銀行，若匯款被中間轉匯行直接自原始匯款金額中扣取國外銀行費用，那麼就會有短收情形發生。

匯款的銀行手續費由誰負擔，可分出三種方式：

OUR (charge originator)

Complete Foreign Bank charges borne by originator

匯款人負擔所有國內及國外銀行的手續費。一般匯款都是要求以此方式來辦理。

SHA

For sharing the Foreign Bank Charges with the beneficiary

匯款人負擔匯款行的手續費，受款人負擔中間轉匯行及解款行的費用。

BEN (charge beneficiary)

Complete Foreign Bank charges borne by beneficiary

受款人負擔所有國內及國外之相關費用，通常匯款行應收之費用會直接從匯款金額中扣除。

因此，買賣雙方在談匯款明細時，除了要清楚說明賣方的銀行戶名、帳號、銀行國際代碼、所屬國家與所在城市之外，若是為OUR這種由匯款人負擔所有銀行手續費的狀況，則還要先問清楚受款人配合的銀行會扣多少手續費，讓匯款人在Invoice金額上另行加上，這樣才不會有匯款短收的問題，也才不會有那種受款人還得要求匯款人補匯那小小差額的情況出現。

Show Time! 換你上場！

翻譯

❶ 可透過銀行匯款、信用卡或郵寄支票來付款。

❷ 可以透過銀行匯款，或是透過PayPal的信用卡付款來支付。

簡答

❶ 請盡你全力，寫出國貿中名詞「匯款、電匯」的各種說法。

❷ 請問「All wire transfer charges should be paid by sender.」是屬於哪一種銀行手續費負擔方式，OUR、SHA或BEN？

來對對答案

翻譯

❶ Payments can be made by bank transfer, credit card or mailed check.

❷ Payment can be carried out by bank transfer or credit card through PayPal.

簡答

❶ remittance, wire transfer, T/T, remittance transfer, international wire, international money transfer

❷ OUR

Chapter 2 搞定訂單

2-4 訂貨確認單

對話 MP3 10

Jimmy: You've reached Formosa Hardware. This is Jimmy Wang, How can I help you?

Vivian: Hi Jimmy, It's Vivian from Western Furniture. Did you receive our order sent yesterday?

Jimmy: Yes, I did. I was just going to call you. On your order, I don't see your billing address. Is that the same as shipping address?

Vivian: Correct. They're the same. So will you Email me your Order Acknowledgement? Our A/C Dept. needs it to make advance payment to you.

Jimmy: I'll Email you today. One thing I need to mention is that we are requiring that all international orders must be confirmed as accurate before shipping. Without those confirmations, orders will be held.

Vivian: Got it! After receiving your Order Acknowledgement, I'll check it, and if any discrepancies are found, I'll advise you immediately by return Email.

Jimmy: Great! That's what I ask for! Thanks for your order and also your cooperation!

背景說明	
人物	Jimmy：原廠業務代表 Vivian：客戶端業務秘書
主題	訂貨確認單的開立與確認： 訂貨確認單：供買方預匯貨款 + 請買方回覆確認

譯文

吉　米：福爾摩沙五金公司，我是王吉米，請問有什麼需要我幫忙的嗎？

薇薇安：嗨，吉米，我是西方傢俱公司的薇薇安，請問您有收到我們昨天發的訂單嗎？

吉　米：有的，我剛才也正要打電話給您呢！在您的訂單中，我沒有看到帳單寄送地址，請問是跟出貨地址一樣嗎？

薇薇安：沒錯，是一樣的。所以您會Email訂貨確認單給我嗎？我們會計部門需要訂貨確認單來安排預先匯款。

吉　米：我會在今天Email給妳。有一件事我要說明一下，我們要求所有的海外訂單都須回覆確認無誤後才會出貨，若沒有確認，訂單就會保留。

薇薇安：瞭解！等收到您的訂貨確認單後，我會檢查檢查，若有什麼不相符的地方，我就會馬上發Email告訴您。

吉　米：很好！這就是我所要求的！謝謝您的訂單，也謝謝您的合作！

說三道四，換句話試試

例一 | 提供訂貨確認單。

例句一 Please find attached our Sales Acknowledgement in relation to your Purchase Order number referenced 00925.

譯文一 請見您採購單單號00925的訂貨確認單如附。

例句二 Thank you for your order. The corresponding confirmation is attached to this Email.

譯文二 謝謝您的訂單，其確認單貼於此Email的附件。

例二 | 提供訂貨確認單，供預付貨款。

例句 Thank you for your order. Attached you will find an order confirmation and one proforma invoice for your prepayment.

譯文 謝謝您的訂單，請見訂貨確認單及形式發票如附，供您預付貨款。

例三 | 提供訂貨確認單，告知供貨訊息。

例句一 Attached please find the order acknowledgement with estimated delivery time.

譯文一 附上訂貨確認單，確認單上列有預估的交貨時間。

例句二 Please find attached your order confirmation. You can find the estimated shipping date on it.

譯文二 請見訂貨確認單如附，上頭有列出預估出貨日。

例四 | 請檢查訂貨確認單上的資訊是否無誤。

例句一 Please check the Order Acknowledgement and, in the event of any discrepancies, please contact us immediately.

譯文一 請檢查此確認單，若有任何不符之處，請馬上跟我們連絡。

例句二 Thank you for your Purchase Order. I am attaching herewith OA #001202 for your review. Please kindly check the order details as attached and let me know immediately if there are any discrepancies.

譯文二 謝謝您的採購訂單，我在此附上訂貨確認單單號001202，供您審閱，還請檢查附件的訂單明細，若有任何不符之處，請立刻通知我。

例句三 Thank you for your order. Attached you will find the order confirmation with the following information:

- ordered products and amounts
- shipping and invoicing addresses

Could you please check whether all the information is correct and let me know if anything should be changed? In this way we can be sure that everything will go smoothly with the shipment.

譯文三 謝謝您的訂單，附件的訂貨確認單有列出了下列訊息：

- 所訂產品、金額
- 出貨地址、發票地址

能否請您檢查所有資訊是否正確？若有需要修改的地方請再告訴我，這樣我們就可確保出貨時順順利利了。

關鍵字急救站—acknowledgement

說到訂貨確認單時，你就有可能一直遇到「acknowledgement」這個字，這字是什麼意思呢？我們先來看看它的英英解釋：「a letter telling you that someone has received something you sent them」，也就是「收件通知、確認通知」的意思，若賣方要跟買方說會處理他的訂單，會再寄確認通知給他，就會這麼說：「I will process your order shortly and send over an acknowledgement.」。接著，我們來看看acknowledgement這個字的字形，它的拼法有兩種，acknowledgement和acknowledgment皆可（差別在一個「e」），是字首ac-加上字根knowledge，再加上名詞字尾-ment，字首ac-也就是加強語氣之用的ak-，我們都知道knowledge是知識，那加上了加強語氣的ac-，怎麼字義就成了「承認、致謝、告知收到（信件等）」呢？來囉！當碰到一個英文字有不同的字義時，就是訓練我們聯想力的最佳時機！請不要輕忽這種聯想的力量，當你將這些不同的字義以有邏輯的推理方式串起來之後，不僅僅是你對這些不同字義會更有印象，同時，你對這個單字的字義也會有更深的體認，換言之，也就是你要感受到字的核心意義了。每一個單字都有它的不同詞性、各種複合字與詞組搭配，當你可以掌握一個單字的核心意義之後，不管這個單字再怎麼變，你就都能推論出它大致的意思，而這樣的功力累積，絕對會培植你的閱讀能力！

好了，說完了小事大真理，我們再回到acknowledgement的幾個不同的字義來聯想……knowledge是知識、知道，當得到了知識，有了一次又一次的強化，才會讓我們「承認」這項知識是真格的，而對方讓我們擁有了這項知識，讓我們知道了某個事實，那真的需要我們誠心「致謝」，若對方是捎來了一封信，發了一項通知，我們收到了、知道了，總是要有禮地回覆對方已收悉，所以acknowledge也有「告知收到（信件等）的字義了，加了名詞字尾-ment的名詞 acknowledgement，意思就是對所來文件表示承認、確認收悉囉！

說完了acknowledgement，我們來看看賣方發了Order Acknowledgement／訂貨確認單給買方後，常會在Email中跟買方說到的一句話：「If any discrepancies are found, please contact us immediately.」，我們在這裡要特別說說的就是「discrepancy」這個字。「discrepancy」是指「a difference between things that should be the same」，本該相同，但卻不同，我們就稱它為「不一致、不符之處」。在字形上，它是來自於拉丁文discrepare，意思為「to sound differently」， 由字首dis-，加上字根crepare而成，dis-表示分離、否定的意思，crepare表示發出咯咯的聲音，就是這種不對勁的聲音，引申出與應該呈現之狀態有不一致的地方。在國貿業務上，則會用到下列這些句子：

- 聲明函裡所述與實際所談有差…
  There are some discrepancies in the Statement.
- 銷售實績與預估有差…
  There is a discrepancy between estimated and actual sales volumes.
- 若訂單內容有任何需說明或任何不相符之處，請跟我們聯絡。
  Please feel free to contact us for any clarifications or discrepancies in the order contents.

下單的「訂單」都叫「order」，而「訂貨確認單」可就不是只有「Order Confirmation」這個說法了，常看到的其他名稱還有：

- Sales Order／銷售訂單
- Order Acknowledgement／訂單確認單
- Sales Acknowledgement／銷售確認單
- P.O. Confirmation／採購訂單確認單
- Confirmation／確認單

由這些詞組不難看出單據名稱的命名邏輯：說到「確認」，就不出這兩個單字：confirmation及 acknowledgement，至於前頭所接的字呢？賣方要確認的標的，為從買方所發出的訂單：Order或 P.O.（Purchase Order／採購單），而對賣方來說，也就是賣方所達成的銷售交易：Sales，所以，訂貨確認單的幾種名稱，就是這幾個字之間排列組合的結果了。

訂貨確認單的功用在於確認訂單已收，所含內容包括如下各要點：

公司資訊	**買方**	訂單處理人員與連絡明細
		收貨方明細
		付款方明細
	賣方	訂單處理人員與連絡明細
下單資料	訂單日期	
	買方訂單單號	
	賣方銷售單單號	
訂單內容	**貨品**	型號
		品名敘述
		規格
		數量
		其他要求（如批號）
	金額	單價
		總價
		運費
		手續費等相關費用
		幣別
出貨	出貨方式	
	預估出貨日	
付款	付款條件	

這些訊息在Proforma Invoice上也都會出現，所以，有些廠商要確認訂單時，也就會直接提供Proforma Invoice給買方，以做為確認訂單明細之用。

接著，我們就實地來看一份訂貨確認單的格式與範例內容囉！

<table>
<tr><td>Sold To:
Energy Health Group
22000 Ellsworth Rd. Ann Arbor, MI, USA</td><td colspan="4">Order Acknowledgement</td></tr>
<tr><td>Sales Order number: 116
Order Date: Oct. 6, 2022
Ref. Number: P14101755
Payment Terms: Net 30</td><td colspan="4">Purchased by: Anke Smith
Email: Ankesmith@energyhealth.com
Your Country: USA
CURRENCY: US Dollars</td></tr>
<tr><td>Ship To:</td><td colspan="4">Bill To:</td></tr>
<tr><td>Energy Health Group
22000 Ellsworth Rd. Ann Arbor, MI, USA</td><td colspan="4">Energy Health Group
22000 Ellsworth Rd. Ann Arbor, MI, USA</td></tr>
<tr><td>Dear Sir / Madam,
Thank you for your order which we confirm as follows:</td><td colspan="4">Please note that items may be shipped separately. Please Email Customer Service at order@aquasystems.com.tw
for the updated shipping date.</td></tr>
<tr><td>ITEM DESCRIPTION</td><td>Est. shipping date</td><td>Spec.</td><td>Qty.</td><td>Unit price</td></tr>
<tr><td>Cat.# SNMW200
Sparkling Natural Mineral Water</td><td>Oct. 12, 2022</td><td>750ml</td><td>10,000</td><td>5</td></tr>
<tr><td colspan="4">Tax:</td><td>0</td></tr>
<tr><td colspan="4">Shipping and handling:</td><td>100</td></tr>
<tr><td colspan="4">TOTAL AMOUNT:</td><td>50,100</td></tr>
<tr><td colspan="5">Invoice number(s) MUST be included with payment. Please pay the full amount of the invoice without deducting bank fees. Customers are responsible for any correspondent and/or intermediary bank transaction fees.</td></tr>
</table>

<table>
<tr><td colspan="1">**買方:**
能量健康集團
美國邁阿密安娜堡埃爾斯沃思路
22000號</td><td colspan="4">**訂貨確認單**</td></tr>
<tr><td>銷售訂單單號：116
訂單日期：2022年10月6日
查詢號碼：P14101755
付款條件：淨30天</td><td colspan="4">採購人：安克．史密斯
Email：Ankesmith@energyhealth.com
國別：美國
幣別：美元</td></tr>
<tr><td>**收貨方明細:**</td><td colspan="4">**付款方明細:**</td></tr>
<tr><td>能量健康集團
美國邁阿密安娜堡埃爾斯沃思路
22000號</td><td colspan="4">能量健康集團
美國邁阿密安娜堡埃爾斯沃思路
22000號</td></tr>
<tr><td>您好，
感謝下單，茲確認如下：</td><td colspan="4">請注意，所訂之品項可能會分批出貨。若欲查詢出貨時間的更新訊息，請與我方客服部門連絡，Email地址為order@aquasystems.com.tw</td></tr>
<tr><td>**品項描述**</td><td>**預估出貨日**</td><td>**規格**</td><td>**數量**</td><td>**單價**</td></tr>
<tr><td>型號 SNMW200
氣泡天然礦泉水</td><td>2022年10月12日</td><td>750ml</td><td>10,000</td><td>5</td></tr>
<tr><td colspan="4">**稅額：**</td><td>0</td></tr>
<tr><td colspan="4">**出貨與手續費：**</td><td>100</td></tr>
<tr><td colspan="4">**總金額：**</td><td>50,100</td></tr>
<tr><td colspan="5">付款時**必須**加註發票號碼，付款金額須為發票全額，銀行手續費不得從中扣除，客戶需自行負擔任何代理銀行及／或中間銀行的交易手續費。</td></tr>
</table>

簡答

❶ 請寫出「訂貨確認單」的三種說法。

❷ 英英解釋：「a letter telling you that someone has received something you sent them」說的是哪一個英文單字呢？

❸ 英英解釋：「a difference between things that should be the same」說的是哪一個英文單字呢？

❹ 訂貨確認單上會有「出貨地址／出貨方明細」與「發票地址／付款方明細」，請問其英文為何？

❺ 訂貨確認單上記載訂單貨品內容處會包含這些欄位，請寫出它們的英文欄位名（請寫其全稱，不縮寫）。

品項描述	預估出貨日	規格	數量	單價

翻譯

❶ 附上訂貨確認單，確認單上列有預估的交貨時間。

Attached please find the ＿＿＿＿＿＿ ＿＿＿＿＿＿ with ＿＿＿＿＿＿ ＿＿＿＿＿＿ ＿＿＿＿＿＿.

❷ 請檢查此確認單，若有任何不符之處，請馬上跟我們連絡。

Please check the Acknowledgement and, in the ＿＿＿＿＿＿ of any ＿＿＿＿＿＿, please contact us immediately.

❸ 若訂單內容有任何需說明或任何不相符之處，請儘管與我們聯絡。

＿＿＿＿＿＿＿＿＿＿＿＿＿＿＿＿＿＿＿＿＿＿＿＿＿＿＿＿＿＿

＿＿＿＿＿＿＿＿＿＿＿＿＿＿＿＿＿＿＿＿＿＿＿＿＿＿＿＿＿＿

來對對答案

簡答

❶ Order Confirmation、Sales Order、Order Acknowledgement、Sales Acknowledgement、P.O. Confirmation或Confirmation（最常見的即為前三種）。

❷ acknowledgement

❸ discrepancy

❹ 有三種表示法：

	出貨地址／出貨方明細	發票地址／付款方明細
1	Ship to	Bill to
2	Shipping address	Billing address
3	Ship-to details	Bill-to details

❺ 欄位名全稱在此：

Item Description	Estimated Shipping Date	Specifications	Quantity	Unit Price

翻譯

❶ order acknowledgement（或其他說法）；estimated delivery time

❷ event；discrepancies

❸ Please feel free to contact us for any clarifications or discrepancies in the order contents.

Chapter 3 篇章簡述

買賣雙方針對出貨所做的討論，會從報價階段就開始，包括價格條件的選擇（也就會決定出運費由哪一方支付）、出貨的方式（一般海空運或快遞等），以及要求、預估的出貨日期。賣方在這些事項的討論上，都可以明白地主動告知買方其偏好的方式，例如：報價均會是根據FOB的價格條件，運費會由買方負擔，而出貨的方式是與像是FedEx這類國際快遞配合，再來，關於出貨的日期，賣方可告知買方預估的出貨日，例如會在收到訂單後五個工作天出貨（shipped 5 working days after receiving your order，或是the lead time is 5 working days）……這些都是買賣雙方在出貨前會確知的訊息。

到了訂立訂單之後，就準備要開展貨物實體移動、出入海關的細節了。其實報關通關的許多細節，大多會由貨物承攬業者打點，賣方要負責的就是確定備齊出口所需的文件，包括出口許可及其他證明，以及追蹤貨的進度，迅速且確實地提供運輸公司與報關行所要的出口相關資料，等貨到了買方國家的海關，若賣方發現進口通關有任何遲延，就要通知買方盡快處理。另外，若所出的貨有儲存溫度要求，則也要主動提醒買方，以確保貨物安全且完好抵達目的地。

在這一章中，我們會先來詳細看看國際貿易的各種出貨方式，瞧瞧在安排出貨上，賣方會跟買方要求哪些資訊？在運費部份，買方會要求賣方提供哪些資訊以供買方估算及比較運費？賣方會要怎麼回覆？這些訊息皆會在3-1為你娓娓道來。

那到了真格出了貨後，就到了真正的「關」卡了：出口通關，賣方主要的工作就是提供文件資料給貨物承攬業者，送交貨品，雖然不用事必躬親，不用親力親為，但也應當瞭解一下出口報關的流程，知道基本的作業內容，這就是3-2要告訴你的知識，要帶你深入淺出地看一下辦理上有時還挺複雜的出口報關作業。

出貨要打通關，出貨文件一定要到！賣方必備的出貨文件包括有空運或海運提單（Air Waybill or B/L）、商業發票（Commercial Invoice）、裝箱單（Packing List）。在3-3中，我們會一一細數這些出貨文件的內容明細，最後也會來看看空運提單與商業發票的實際格式與內容，讓你瞭解得又通又透！

Chapter 3 搞定出貨

3-1 出貨方式

Evelyn: Hello, Evelyn speaking. How may I help you?

Vincent: Hi, Evelyn, This is Vincent from Laurel Cosmetics in USA. I'm calling to check when our ordered catalog will be ready for despatch?

Evelyn: We hope to ship next week. By the way, I do need your confirmation of your desired method of shipment. Would you like me to solicit a FedEx international Economy Air or Sea Freight quote for you?

Vincent: No, thanks. We have our preferred carrier to make shipping arrangement for us. And we'll use air freight because we need these catalogs urgently for the upcoming exhibition. Please just tell me the pickup address and also the packaging box details for us to check its freight, like box dimensions and gross weight.

Evelyn: No problem. I'll gather these details and inform you by Email. Please also let me know the contact information of your air freight forwarder should this become necessary.

Vincent: Okay, I'll also send you an Email later.

Evelyn: Thanks!

背景說明	
人物	Evelyn：原廠行銷部門人員 Vincent：經銷商產品專員
主題	確認出貨方式，買方有配合的貨運公司： 空運、FedEx空海運、貨運前置安排作業

譯文

艾芙琳：哈囉，我是艾芙琳，有什麼我可以效勞的地方嗎？

文　生：嗨，艾芙琳，我是美國桂冠化妝品公司的文生，我想要跟您確認一下我們訂的型錄什麼時候可以出貨呢？

艾芙琳：我們希望是下星期可以出貨，對了，我需要跟您確認一下出貨方式，您要我們請聯邦快遞報給您經濟型空運或海運的運費嗎？

文　生：不用了，謝謝，我們有自己配合的貨運公司會幫我們安排出貨，還有，我們會走空運出貨，因為展覽時間快要到了，我們急著要這些型錄。就請告訴我們取貨的地址，還有包裝裝箱的相關明細，像是箱子的尺寸和毛重，這樣我們就可以去查查運費會是多少。

艾芙琳：沒問題，我會整理這些細節後，發個Email告訴您。請也讓我們知道這家空運公司的連絡資料，若有必要我們就可連絡得上。

文　生：好的，我待會兒也會發個Email給您。

艾芙琳：謝謝了！

說三道四，換句話試試

例一 | 郵寄出貨

例句 Orders below 1,000 USD can be shipped by air mail post but such shipment is not possible to track and can take up to 20 working days to arrive.

譯文 金額小於1,000美元的訂單可安排郵寄出貨，不過這種出貨無法追蹤，而且最久會需要20天才能到貨。

例二 | 走海運

例句 Sea freight may be your preferred choice of transportation since this order consists of several large items.

譯文 海運可能會是您想要安排的運輸方式，因為這個訂單裡有好幾個大型的貨物。

例三 | 走快遞

例句 Since middle of November, we are offering courier shipment by FedEx for 60 USD world-wide.

譯文 從十一月開始，對於全球的出貨我們都可安排走FedEx快遞，而且運費皆為60美元。

例四 | 出貨方式任你選

例句 We can arrange shipments by using courier service, such as Fedex or DHL. We also ship by air freight and sea freight but will advise you of the best options at time of order.

譯文 我們可以安排快遞出貨，例如走FedEx或DHL，此外，我們也可安排空運或海運出貨，不過會在您下單時才能跟您建議哪一種出貨方式最適合。

例五 | 特殊出貨

例句 This product must be stored and shipped frozen. Do you have any experience importing frozen goods from Taiwan? We would need you to advise us what means of transport to use for the shipment, FedEx or air freight.

譯文 此產品必須冷凍儲存與出貨，您有從台灣進口冷凍貨物的經驗嗎？我們得請您告知要以什麼運送方式出貨，要走FedEx或是一般空運。

例六 | 運費

例句 We will let you know the dimensions and gross weight of the shipment for you to check the estimated freight with your carrier.

譯文 我們會告訴您出貨貨物的尺寸和毛重，讓您與您的貨運公司查核預估運費。

關鍵字急救站

買賣雙方在討論出貨相關事宜時，不論如何，一定會用到這幾個單字：ship, shipping, shipment, deliver, delivering, delivery, despatch，似乎意思都是…出貨？沒錯，都可說出貨，但其中還是有些許的差異，我們就選這幾個字的名詞來探探異同囉！

- shipment [ˋʃɪpmənt]
  n. 出貨；運輸 the process of taking goods from one place to another
- delivery [dɪˋlɪvərɪ]
  n. 送貨 the process of bringing goods or letters to a place
- despatch [dɪˋspætʃ] = dispatch（英式拼法）
  n. 派遣；發送； the sending of someone or something to a destination or for a purpose

看了英英解釋後，你應該可以馬上叫出despatch跟其他兩個字的不同：despatch處理的標的物還可以是someone，可以對到不是東西的人（我…我可沒在罵人喔）！而在對到物的時候，despatch又跟其他兩個字差不多了，像是若要跟買方說等跑完書面作業，文件送到倉庫後，即可派送訂單，就可說：「I will get the paperwork to our warehouse asap to dispatch your order.」。

那shipment 跟delivery就是完全一樣的嗎？喔不！世界沒這麼單純啊！雖說一般人常將shipment跟delivery都說成「出貨」，但出貨要讓貨從一地移往另一地，會有在途的期間，所以光出貨這事就可劃分為兩個端點：

➡ Shipping date／出貨日: It tells that the product has been sent on the given date.

➡ delivering date／到貨日: It is the date when the customer receives the product.

另外，**shipment** 跟 **delivery**的差異還有一說，說到**shipment**用在較小貨品的出貨上，而**delivery**則是指較大貨品的出貨。不過，實務上說到出貨，倒也沒盯住這一點來做區分。

接著，我們就來看看這些字的用法囉！

➡ shipping documents／出貨文件

➡ shipping notice／出貨通知單

➡ shipping agent ／貨運代理商

➡ Please be advised we will make shipment of your order the following Monday.
在此通知您，我們將在下個星期一為您的訂單安排出貨。

➡ For up-to-date information about the delivery time for your order, please visit your account on our website.
請到我們的網站查看您的帳戶，就能夠查到關於您訂單到貨時間的最新消息。

術語直達車，專業補給站

在談妥了訂單，也正式下單了之後，客戶會關心的頭一件事就是：何時出貨。那麼貨會是怎麼出的呢？出貨方式的選擇，取決於成本考量與時間要求，若貨不急，也就沒必要選用最快的運輸方式，若貨很急，運費再貴也要咬牙花下去的啊！

國際貿易的出貨方式不外乎海運與空運，如果以承接貨物、辦理通關的中介角色來區分，則可分出下列這幾種方式：

- 一般海空運
- 快遞公司海空運
- 郵局快捷、一般航空、水陸、陸空聯運

上述的幾種不同空運方式，以速度來看，快遞空運快過一般空運，一般空運與郵局快捷差不多，郵局快捷又快過郵局的一般航空。而在運費上，要快當然索價就高的囉！

出貨方式不同，連處理的相關人／公司的稱呼也就各不相同，各有習慣。請來認一認這些你常掛在口中的詞兒吧：

出貨方式	賣方	運送方	買方
一般海空運	Shipper／Consignor (出貨人／託運人)	Carrier (運送人) Forwarder (貨代)	Consignee (受託人／收件人)
快遞公司	Sender (寄件人)	Courier (快遞公司)	Recipient (收件人)
郵局	From (從／寄件人)	Post office (郵局)	To (到／收件人)

在安排出貨上（coordinate shipments），賣方會要跟買方要求下列資訊：

- Shipping address, phone number and contact person
- Preferred carrier and shipping method

買方須提供收貨人的基本資訊給賣方，包括出貨地址，電話與連絡人這些基本身家資料，以及對出貨方式的要求，看是要走一般海空運、走快遞，還是透過郵局來寄貨。

此外，賣方還會問買方有沒有什麼偏好的貨物承攬業務代理公司（簡稱貨代），如有，會請買方將賣方的連絡資料給貨代，請貨代主動來跟賣方連絡，也會提醒買方得要負責安排所有文書作業與取貨事宜，英文可以這麼寫：

- If using a Freight forwarder, please provide them with our contact information.
- If you choose a freight forwarder option, please note that you will be responsible for coordinating all paperwork and scheduling the pickup.

若買方要求走快遞空運，賣方會問要配合的是哪一家快遞公司，如買方要自付運費，那就得要知道買方的快遞公司帳號，讓賣方在出貨提單上填寫買方的帳號，那麼快遞公司就會以與買方簽約的條件來計算運費，跟買方收取。賣方所要做的出貨準備工作包括包裝、準備託運文件，也就是填寫空運提單（Air Waybill）或海運提單（Bill of Lading）、商業發票

（Commercial Invoice）、裝箱單（Packing List），隨貨附上，接著再連絡快遞公司到寄件人處收件（Pickup），完成交寄作業。貨一送出，接著就是辦理出口通關、進口清關（clearance）的作業了，待過了這國內外海關的這一關，買方就可等著貨物送上門了（delivery）。

在出貨的運費部份，買方要決定選擇哪一家公司為貨代時，通常會要求賣方提供下列這個資訊，以供估算及比較運費：

➡ Please let us know the gross weight and dimensions of the shipment.
請告知此批貨的毛重與尺寸。

要計算海運與空運的運費，就得要知道貨物的體積與重量了，大原則是這樣的：若體積大於重量，則以體積來估算運費，若是重量大於體積，則以重量估算。一般來說，海運都以體積來計，即使有時重量大於體積，但最終運費差異上也不會太大。空運一般都以重量來計，不過有些貨品雖輕但很佔空間，材積重大於重量，這時就會以材積重來計費了。賣方要回覆買方重量與計算材積的尺寸這些數據時，可以這麼說：
例：出貨共有8箱，每箱尺寸為18 x 16 x 11（吋），重約20磅

➡ There will be a total of 8 boxes. Each box would be around 20 pounds with the dimensions of 18 x 16 x 11 (inches).

➡ This shipment will contain 8 boxes.
Dimensions (inches): 18 x 16 x 11
Weight (pounds per box): Approx. 20

➡ Packaging:

8 boxes - Each box measures 18 x 16 x 11 inches and weighs approx. 20 pounds.

買方估算了運費，決定了貨代之後，就可連絡貨代前往賣方處取貨了。

若要透過郵局來出貨，中華郵政提供了一般國際包裹及國際快捷的服務，國際快捷亦稱EMS（International Express Mail Service），寄件人須填寫五聯單與商業發票，至郵局窗口辦理，出貨後有EMS號碼可追蹤交寄的貨物。不過，郵局寄送服務限制較多，有禁寄物品的規定，還有尺寸與重量限制，在重量部分，依寄送國家不同，限重20公斤或30公斤。有時還會看到美國廠商通知透過USPS出貨，乍看會以為是UPS美商優比速快遞（United Parcel Service of America）， 其實指的是USPS（United States Postal Service），也就是美國郵局哩。

出貨的方式有好幾種，可選擇的運輸公司也有個好幾家，趁早找到固定配合的貨運業者、代理人，談個好費率，在配合上也能愈來愈有默契。如此一來，出貨安排就不難，出貨就能出得順了，而要是真出了什麼問題，也比較好控管情勢，打通關卡，讓出貨事繼續順下去囉！

填空

❶ 不同出貨方式所稱呼的各方人馬：

出貨方式	賣方	運送方	買方
一般海空運	________ 出貨人／託運人	________ 運送人 ________ 貨代	________ 受託人／收件人
快遞公司	________ 寄件人	________ 快遞公司	________ 收件人
郵局	________ 從、寄件人	________ 郵局	________ 到／收件人

❷ 請寫出「出貨」的三個大同小異的名詞：

出貨、運輸：________

送貨：________

派遣、發送：________

❸ 出貨日：________ ________

到貨日：________ ________

翻譯

❶ 我們可以安排快遞出貨，例如走FedEx或DHL，此外，我們也可安排空運或海運出貨，不過會是在您下單時再跟您建議哪一種出貨方式最適合。

We can arrange shipments by using ________ ________ such as FedEx or DHL. We also ship by ________ ________ and ________ ________ but will advise you of the best options at time of order.

❷ 我們得請您告知要以什麼運送方式出貨，要走FedEx或是一般空運。

____

____

❸ 海運可能會是您偏好的運輸方式，因為這個訂單裡有好幾個大型貨物。

____

____

來對對答案

填空

❶ 請往前四頁找答案囉。

❷ shipment；delivery；despatch

❸ shipping date；delivering date

翻譯

❶ courier service；air freight；sea freight

❷ We would need you to advise us what means of transport to use for the shipment, Fedex or air freight.

❸ Sea freight may be your preferred choice of transportation since this order consists of several large items.

Chapter 3 搞定出貨

3-2 出貨通關

Celine: Hi, I'd like to speak to Benny Clapton, please.

Benny: Benny speaking. How can I help you?

Celine: This is Celine at Hallmark Antigen Enterprises. I just received a customs clearance delay notification from FedEx regarding the shipment that we arranged for you two days ago. Can you please provide the information they require so as to have the package released asap?

Benny: That's the issue we're working on right now. Your call just came at the right time! We're requested by our customs to provide its invoice and also the product's data sheet.

Celine: The Invoice was already included on the outside of the package. But it's okay. I can Email the Invoice to you together with the data sheet later.

Benny: That's great. Thanks for your help.

Celine: I'd like to remind you that the product is temperature sensitive, and therefore, please make sure that the parcel is being stored in the appropriate temperature.

Benny: Thanks for your reminder. We'll pay attention to that!

背景說明	
人物	Celine：原廠業務秘書 Benny：經銷商業務助理
主題	出貨通關延遲，須補文件給海關： 補Invoice、產品說明書，另原廠提醒要注意來貨儲存溫度

譯文

席琳： 嗨，麻煩請找班尼．克萊普頓。

班尼： 我就是，有什麼需要我幫忙的嗎？

席琳： 我是標誌抗原企業的席琳，我們兩天前有出貨給您，但剛收到了FedEx發來的通關延遲通知，還請您提供海關所需資訊，好讓貨能盡快放行。

班尼： 我們現在就在處理這事，您的電話來得正好！海關要求我們提供發票以及產品的說明書。

席琳： 我們有將發票附在貨的外包裝上了，不過沒關係，我等一下就可以Email給您，也會另外附上說明書。

班尼： 太好了，謝謝您的幫忙。

席琳： 我想要提醒您一件事，這次所出的貨屬於對溫度敏感的產品，所以要請您務必確認要將包裹儲存在適當的溫度下。

班尼： 謝謝您的提醒，我們會注意的！

說三道四，換句話試試

例一 | 賣方通知貨卡關，請買方處理。

例句一 It seems that the parcel is undergoing a clearance delay. Could you contact your local FedEx to see what is required to speed up the custom clearance?

譯文一 此包裹似乎通關有遲延，是否您能與您當地的FedEx連絡，看看需要怎麼做才能加速清關？

例句二 We noticed that a customs clearance delay is caused because some documents are missing. Please handle this issue asap and let us know if we can be of any assistance.

譯文二 我們注意到通關有遲延的狀況，原因是缺了些文件。請盡速處理此事，若有需要我們提供協助的地方，也請讓我們知道。

例二 | 買方通知貨抽驗，請賣方補資料。

例句一 This shipment is selected for inspection by our FDA. FDA requests the manufacturer to provide the Certificate of Analysis inclusive of the words “For Research Use Only”. Please help and Email to us asap.

譯文一 FDA抽驗抽到我們這次的來貨，要求廠商提供的產品分析報告上要有「僅供研究使用」的字樣，請幫忙盡快Email給我們。

例句二 This shipment needs to go through customs inspection. Since total 4 boxes were shipped, we're requested to submit the Packing List which contains box no. Please revise your Packing List accordingly and Email to me today.

譯文二 這次的來貨得經過海關檢驗，因為總共出了四箱，海關要求我們提供的裝箱單要列有箱號，還請您依照要求修改裝箱單，並在今天Email給我。

例三 | 賣方提供了通關所需資料，問還有無其他問題。

例句 Attached please find the signed invoices requested by your customs. Please let me know if you still have any difficulties with customs clearance.

譯文 附上您海關所要求的含簽名的發票，如果通關上還有什麼困難，再請告訴我們。

例四 | 述明與通關相關的責任歸屬。

例句一 The buyer should be responsible for all import customs clearance formalities plus import duties and taxes.

譯文一 買方應負責辦理所有通關手續，並支付進口相關稅負。

例句二 Please be advised that the buyer is responsible for customs clearance costs and the payment of duties and taxes.

譯文二 在此通知您，買方應負擔通關成本，並支付進口相關稅負。

關鍵字急救站

賣方出了貨品，貨要先經過出口國家的海關出關，之後再到進口國家的海關通關、放行，而在海關所辦理的通關（或稱清關），英文就是customs clearance。我們先來看看customs／海關這個字的英英解釋：「the place at a port, airport, or border where officials check the goods that people are bringing into a country are legal, and whether they should pay customs duties」。

所以不管是安排空運飛到機場，或是走海運駛入港口，這個負責控管貨物進出國境的政府主管機關就稱為「海關」，而「古者境上為關」，所謂的關，指的就是進出國門的關口，所以海關所收之稅叫關稅，即是customs duty，海關申報表就是customs declaration。另外，在出貨文件的發票中，最常用到customs這字的是說到所列金額是僅供海關參考，像是要寄送樣品時，就會在發票金額下方這麼加註：「VALUE FOR CUSTOMS PURPOSES ONLY」。

若是有必要退貨給賣方，則賣方會要求買方在商業發票上特別說明一下：

➡ Please note on Commercial Invoice: No Sale or transaction has occurred; value stated is for customs purposes only.
請在商業發票上加註：並未有任何銷售或交易行為發生；所列金額僅供海關參考。

古時所說的「關之賦」就是關稅。關稅的英文有二：「customs」與「tariff」，其英英解釋分別如下所示，看完你就會放心地說：「customs和tariff 兩個字真的一樣哩」！

- customs: the taxes that you pay on goods that you bring into a country
- tariff: a tax that a government charges on goods that enter or leave their country

看完了海關與關稅，再來我們看看這個關要「通」要「清」所說的clearance，它是指「the clearing of a person or ship by customs」，其中的動名詞clearing來自於動詞clear，字義就是「為貨物結關、通過（海關等）」。

最後，我們再一塊兒來看看會說到customs、customs clearance的其他例句囉！

- We always declare the true value of our products in the commercial invoice for customs clearance.
  我們在商業發票上所列的都是產品的真實金額，供辦理清關。
- We'll need these documents so as to clear the goods from our customs.
  我們清關時就得要用到這些文件。
- I have signed the invoices as requested and, if you have any issues with customs, please advise so we may adjust accordingly.
  我們已依您要求在發票上簽名了，若是海關還有其他要求，請告訴我們，我們可配合調整。

術語直達車，專業補給站

賣方要辦理出口通關時，在作業上多會找配合的貨物承攬業者來處理，因此，賣方主要的工作就是提供文件資料與送交貨品，而所對應的窗口也就是貨物承攬業者，不用親力親為，不用花上許多力氣與海關接洽，但我們還是應當知道一下出口報關的流程，知道基本的作業內容，這樣賣方在與貨物承攬業者交談時，也才能更懂得怎麼詢問，而在向買方説明狀況時，也才比較能知道自己説的到底是什麼東西呢！所以，在此就讓我們簡要地來看一下辦理上有時還挺複雜的出口報關作業囉！

1. 賣方提供出貨通知給貨物承攬業者

賣方確定了出口貨物與受託人（consignee）相關明細之後，即可連絡貨務承攬業者，讓其洽訂艙位（booking），完成訂位後，貨物承攬業者就會核發訂艙通知單（Booking Note）或裝船通知單（Shipping Advice）給賣方。

2. 賣方提供報關所需文件（Submission of documentation）

賣方備妥包括發票（Invoice）和裝箱單（Packing List）這些基本文件，供貨物承攬業者直接進行報關作業，或經由報關業者來報關。

3. 出口報關（Export customs declaration）作業

報關業者收到出口廠商提供的文件和裝船通知單後，將其內容製做成出口報單，透過電子資料交換（Electronic Data Interchange，簡稱EDI）的傳輸方式，經由關貿網路（Trade-Van，簡稱T/V；Van為頭字語：

Value Added Network）將資料傳至海關，經專家系統（Expert System）比對，自動篩選通關方式，分出不同報單。海關電腦會篩選出C1、C2、C3報單：

- C1免審免驗（Free of Paper and Cargo）

  C1為免審文件、免驗貨物的報單，即予放行（Authorizing release），之後經電腦抽核分為補送報單審核及免補送報單審核（即無紙化C1NP、無紙化通關，免填出口報單）。

- C2文件審核（Document Scrutiny）

  報關業者須於規定時間內，向海關補送書面報單及相關文件，經海關收單及完成分估（分類估價計稅）作業後，方予放行。

- C3貨物查驗（Cargo Examination）

  報關業者須於規定時間內，向海關補送書面報單及相關文件，經海關收單，並查驗貨物，完成分估作業後，方予放行。

4. 賣方接收出貨文件（Receiving shipping documents）

待出口報關放行後，貨運承攬業者即可製作提單，將文件彙總提供給賣方，文件包括：

Air waybill / B/L 空運或海運提單

Commercial Invoice 商業發票

Packing List 裝箱單

我們在前一個單元也有說到了貨運承攬業者這個角色，在此我們就來為其負責的角色與任務再多加說明一下囉！

- 貨運承攬業者主要的工作在於安排貨物的進出口運送，以將不同委託人的零星貨物併成整櫃，再交給實際的運輸業者運送，包括航空公司與船公司。
- 貨運承攬業者所扮演的是介於委託的出口廠商與運輸業者之間的中介角色，以自己的名義代為處理進出口貨物的運輸業務。
- 有些貨運承攬業者還會負責報關業務，不一定會交由報關行 (Customs Broker) 負責，尤其是空運業者，因為時間有限，通常都會由空運承攬業者同時負責報關作業。

最後，就讓我們一起來看看出口報關的流程圖，若你對出口報關還有些糊糊不清楚的地方，看了流程圖之後，一定會心中清明、腦袋清楚的啦！

出口報關流程：

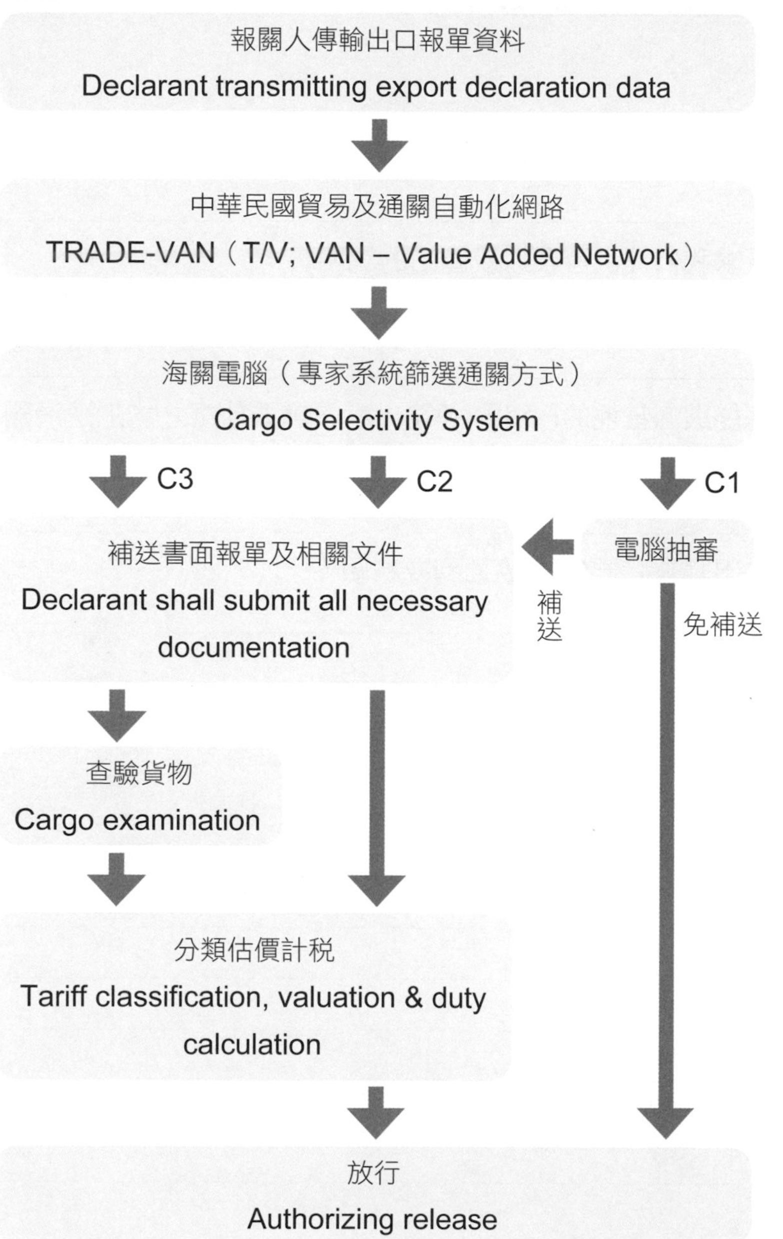
報關人傳輸出口報單資料
Declarant transmitting export declaration data
中華民國貿易及通關自動化網路
TRADE-VAN（T/V; VAN – Value Added Network）
海關電腦（專家系統篩選通關方式）
Cargo Selectivity System
C3
C2
C1
補送書面報單及相關文件
Declarant shall submit all necessary documentation
電腦抽審
補送
免補送
查驗貨物
Cargo examination
分類估價計稅
Tariff classification, valuation & duty calculation
放行
Authorizing release

Show Time! 換你上場！

翻譯

❶ 我剛收到了FedEx發來的通關延遲通知。

❷ 海關要求我們提供發票以及產品的説明書。

❸ 是否您能與您當地的FedEx連絡，看看需要什麼才能加速清關？

❹ 這次的出貨被我們的FDA抽到要檢驗。

術語翻譯

No.	中文	英文
1	裝箱單	
2	出口報關	
3	C1免審免驗	
4	C2文件審核	
5	C3貨物查驗	
6	放行	

翻譯

❶ I just received a customs clearance delay notification from FedEx.

❷ We're requested by our customs to provide the Invoice and also the product data sheet.

❸ Could you contact your local FedEx to see what is required to speed up the custom clearance?

❹ This shipment is selected for inspection by our FDA.

術語翻譯

❶ Packing List

❷ export customs declaration

❸ C1 Free of Paper and Cargo

❹ C2 Document Scrutiny

❺ C3 Cargo Examination

❻ authorizing release

Chapter 3 搞定出貨

3-3 出貨文件

對話 MP3 13

Rene: Hello, Jasmine. It's Rene from Amber Biomedical.

Jasmine: Hi, Renee. Nice to hear from you. How've you been?

Rene: Not bad. I'm calling to check whether you've shipped our order.

Jasmine: Yes, we shipped it out yesterday.

Rene: Great! Please Email me a full set of its shipping documents.

Jasmine: Not a problem. I'll Email the Air Waybill, Invoice and also Packing List to you later.

Rene: Did you pack the product's data sheet with the shipment?

Jasmine: No... Did you need it for customs clearance?

Rene: If this shipment is selected for cargo examination, we'll need to submit its data sheet. So we'd like to have it ready in case it's requested.

Jasmine: I see. I'll Email along with shipping documents. If you need any other documents for clearing the goods from customs, please just let me know.

Rene: Okay. Thanks!

背景說明	
人物	Rene：經銷商業務秘書 Jasmine：原廠客服人員
主題	問出貨，要求提供出貨文件： 已出貨，買方要求提單、商業發票、裝箱單 & 產品說明書

譯文

雷　內：哈囉，潔思敏，我是琥珀生醫公司的雷內。

潔思敏：嗨，雷尼，很高興您打電話來，您好嗎？

雷　內：還不錯，我打電話來是想問問我們的訂單出貨了嗎？

潔思敏：有，我們昨天出貨了。

雷　內：太棒了！請Email整套出貨文件給我。

潔思敏：沒問題，稍後就會Email提單、商業發票和裝箱單給您。

雷　內：請問你出貨時有附上產品的說明書嗎？

潔思敏：沒有耶……你們通關需要說明書嗎？

雷　內：如果貨物被抽驗，就得提供說明書，所以我們想先備好，有需要時就用得上。

潔思敏：瞭解，我會跟著出貨文件一起Email給您，若是您還有需要什麼其他文件來辦理清關，請告訴我一聲。

雷　內：好的，謝謝！

例一 | 賣方通知出貨，送上出貨文件。

例句一 Please see below for your shipping details and the attached for the AWB, Packing Slip and Commercial Invoices.

譯文一 請見出貨明細如下，以及提單、裝箱單、商業發票如附。

例句二 The package has been shipped via FedEx with tracking # 773500960925. Enclosed please find the documents for customs clearance.

譯文二 此包裹已透過FedEx出貨，追蹤號碼為773500960925，在此附上通關需要的文件如附。

例句三 We're pleased to inform you that your purchase order No. 10931202 has been shipped already. You can follow the shipment by logging into www.fedex.com and insert the tracking number to find where your package is. We also attach a copy of the documents that were sent along with the shipment.

譯文三 我們很高興地要來通知您採購單單號10931202的貨已出，您可登入www.fedex.com，輸入追蹤號碼，就可知道包裹在途中何處。我們也在此附上隨貨寄出的文件。

例二 | 賣方應買方要求，提供文件

1. 要發票

例句 Per your request, please find a copy of your invoice attached for customs clearance. If there's any additional paperwork aside from the Invoice that your customs requires, please just let me know.

譯文 依您所要求，在此附上發票影本供您通關之用，若除了發票之外還有需要其他的文件，就請儘管告訴我們。

2. 要整套出貨文件

例句 Your order has been shipped out via FedEx AWB #6107-9687-6992 and should arrive shortly. Per your request, please find the complete set of shipping documents attached. The original ones will be packed together with the shipment.

譯文 您的訂單已出貨，聯邦快遞提單號碼為6107-9687-6992，應很快就可送達。依您所要求，在此附上一份完整的出貨文件，正本將會隨貨寄出。

3. 品名要與出貨文件所列相符

例句 Please find attached the QC data sheets for the two products included on Order #15505. As requested, I have modified the QC data sheet so the product name will match the invoice and other shipping documents.

譯文 在此附上訂單單號15505中兩項產品的品管說明書，依您的要求，我已修改了品管說明書，使其產品品名與發票及其他出貨文件所列相符。

關鍵字急救站

要說出貨文件的關鍵字，我們就該來好好談談一種必要的出貨文件：提單！提單的英文名稱我們常看、常寫，還滿簡單的啊！但是，真的簡單嗎？沒想過都覺得簡單，開始想才發現還有些地方不簡單，等想完了之後，因為知識水平提升了，就又變為簡單了呢！

空運與海運提單的名稱不同，分別是：

- 空運提單：Air WayBill
- 海運提單：Bill of Lading

兩個詞都有「bill」，那我們當然就要先來看看這個「比爾」囉！「bill」不就是「帳單」嗎？若真要說到帳單，那不是出貨文件中的「發票」所要負責的角色嗎？是啊！所以，得另求高明，繼續尋找最恰當的字義。在你查了英英解釋之後，反而會一下子找不到合適的解釋，但請再仔細看看，你會發現最貼近、最留有空間可供沿用的解釋是這麼說明的：「a list of events and performances at a concert, show etc.」，所以是…節目單？喔不，那個「etc.」可重要了，有了這個開闊的視野，我們可歸類說明提單為一份「清單」，那是什麼清單呢？我們從海運派來看，海運提單是「Bill of Lading」，是一份「lading」的清單，而「lade」的意思是「put cargo on board (a ship)」，也就是裝載、裝船的意思，所以，「Bill of Lading」這個語詞的意思就是：「a detailed list of a ship's cargo in the form of a receipt given by the master of the ship to the person consigning the goods」，說著這份bill就是一份詳細的清單，是船上所裝貨物的清單，用來做為送交貨物的收據。

說完了海運派，我們來看看它的另一個門派–空運派。我們似乎有看過空運提單這麼寫：「Air WayBill」，也有看過這樣寫的：「Airway Bill」，那way跟air併，或是跟bill併，意思一樣嗎？「airway」這個字的字義包括有（礦井的）風道、（肺的）氣道、（麻醉時用的）導氣管，咦？都跟出貨很沒關係哩！是的，我們談國貿，它的字義一定要合乎這個情境脈絡，所以請查字典時繼續往下找，你會看到「航空路線、航空公司」的解釋，這才是對的，若寫成「Airway Bill」，可兜得上的意思為「航空公司的單據」。那「waybill」是什麼意思呢？它的英英解釋為「a list of passengers or goods being carried on a vehicle」，意思為運輸工具所載送貨物的清單，這可是比「Airway Bill」／「航空公司的單據」強多了、完整多了，所以，當你下回要寫這個文件的名稱，請記得要寫為「Air Waybill」喔！

術語直達車，專業補給站

既然要說出貨文件，那我們就狠狠地來把幾份基本且必須的出貨文件看個透吧！

一、提單

❶ 海運提單／Bill of Lading（B/L）

提單是出口廠商出貨時所填的單據，通常是由貨運承攬業者提供，是出口廠商與貨運承攬業者之間對託運貨物所立下的合約，記載相關的權利義務，為貨權的憑證。貨運承攬業者簽收後，會將一份提單給出口廠商，做為收據，此時，對貨物的責任即從出口廠商移轉至貨運承攬業者，直到將貨交給進口廠商為止。如有任何的索賠情事發生，提單即為必要的單據。

❷ 空運提單／Air Waybill（AWB）

空運提單項下分有兩種提單：

- 空運主提單（Master Air Waybill；Master AWB）
- 空運分提單（House Air Waybill；House AWB）

空運主提單為由航空公司所出具，提單號碼的前三碼為三個阿拉伯數字，此三碼數字即為航空公司的三位數識別代碼，例如長榮航空公司的三位代碼為695，代碼後頭再跟著航空公司所編不超過八位數數字的貨號及帳號流水號，例如695-87994742。

空運分提單為由航空貨運承攬公司所出具，提單號碼的前三碼為該貨運承攬公司的英文字母代號碼，代碼後頭再跟著其自編的流水號，例如NEU-9732 9864。因為航空貨運承攬公司本身並不是實際的運送人，也未必是實際運送人的代理人，所以其所出具發行的分提單僅為出口廠商與航空貨運承攬公司之間的運送契約，如有任何的索賠情事發生，出口廠商只能向航空貨運承攬公司主張權利，而不能直接對航空公司主張任何的權利。

出口廠商所出的貨物，在主提單與分提單中的「份量」並不相同。空運主提單是個大單，裡頭包括一件以上的集裝貨物，是多件貨運承攬業者所受託之貨物的彙集。而空運分提單則有單一性，每件集裝貨物都會有一個分提單單號，貨運承攬業者以此來分辨各件不同的受託貨物。

有人主張空運提單應稱為「空運運單」，所持的理由即為海運提單是代表貨物所有權的憑證，以交出提單作為提貨的條件，但空運提單並不是，就算持有空運提單也不代表可以對貨物要求所有權，亦不能通過背書來進行轉讓，所以僅為貨運運輸的單據，故應稱空運運單。

二、商業發票／Commercial Invoice

商業發票在國貿上為具有效力的買賣證明單據，上頭會載明該批運送貨物的規格、型號、包裝與金額等資訊，由出口廠商在出貨時出具，提供給進口廠商做為清關的憑證。

三、裝箱單／Packing List、Packing Slip

裝箱單是由出口廠商在出貨時出具，是貨品內容的說明憑證，與商業發票的內容大致相同，差別僅在商業發票會載明金額，而裝箱單則不列金額。

接著，我們就來實地看看空運提單與商業發票的格式與內容囉！

出貨文件 一 空運提單 **Air Waybill** 範例

297 8252 1000

Shipper's Name and Address	Shipper's Account Number	Not Negotiable Air Waybill issued by Bright Freight Inc.
BioChem Science, Inc. 37F, No. 337, Ruiguang Rd., Neihu Dist., Taipei, 114, Taiwan		Copies 1, 2 and 3 of this Air Waybill are originals and have the same validity
Consignee's Name and Address Chemport Corp. 1000 xxxx Ave Chino, CA 91710, USA	Consignee's Account Number	It is agreed that the goods declared herein are accepted in apparent good order and condition (except as noted) for carriage SUBJECT TO THE CONDITIONS OF CONTRACT ON THE REVERSE HEREOF. ALL GOODS MAY BE CARRIED BY ANY OTHER MEANS INCLUDING ROAD OR ANY OTHER CARRIER UNLESS SPECIFIC CONTRARY INSTRUCTIONS ARE GIVEN HEREON BY THE SHIPPER, AND SHIPPER AGREES THAT THE SHIPMENT MAY BE CARRIED VIA INTERMEDIATE STOPPING PLACES WHICH THE CARRIER DEEMS APPROPRIATE. THE SHIPPER'S ATTENTION IS DRAWN TO THE NOTICE CONCERNING CARRIER'S LIMITATION OF LIABILITY. Shipper may increase such limitation of liability by declaring a higher value for carriage and paying a supplemental charge if required
Issuing Carrier's Agent Name and City Bright Freight Inc. 5F, No. 5, Yong Ji Rd., Xin Yi District 110 Taipei Taiwan		Accounting Information Hoping Freight Ltd. 1500, XXXX AVE. CHINO, CA 91710 Tel: 909-548-6000 Fax: 909-548-6111 Attn.: David
Agent's IATA Code 01-1-9000/0010	Account No.	

Airport of Departure (Addr. of First Carrier) and Requested Routing TAOYUAN Airport							Reference Number		Optional Shipping Information			
To TPE	By First Carrier CI5100/10	Routing and Destination	to	by	to	by	Currency US$	CHGS	WT/VAL PPD | COLL	Other PPD | COLL	Declared Value for Carriage	Declared Value for Customs

Airport of Destination CHINO Airport	Requested Flight/Date 12/11/2022		Amount of Insurance	INSURANCE - If carrier offers insurance, and such insurance is requested in accordance with the conditions thereof, indicate amount to be insured in figures in box marked "Amount of Insurance".

Handling Information	SCI

No. of Pieces RCP	Gross Weight	kg lb.	Rate Class Commodity Item No	Chargeable Weight	Rate/ Charge	Total	Nature and Quantity of Goods (incl. Dimensions or Volume)
2 CTNS	50L 25K		50L 25K				Lab Research Reagents (Research Purpose Only) "Freight Prepaid" This shipment is subject to inspection Dry ice needs special handing Box 1- PLS keep frozen

Prepaid	Weight Charge	Collect	Other Charges Prepaid
Valuation Charge			
Tax			I hereby certify that the particulars on the face hereof are correct and that insofar as any part of the consignment contains dangerous goods. I hereby certify that the contents of this consignment are fully and accurately described above by proper shipping name and are classified, packaged, marked and labeled, and in proper condition for carriage by air according to applicable national governmental regulations. BIOCHEM SCIENCE
Total Other Charges Due Agent			
Total Other Charges Due Carrier			
			Signature of Shipper or his Agent
Total Prepaid	Total Collect		
Currency Conversion Rate	CC Charges in Dest Currency		AS AGENT FOR 12/11/2022 LOS TAIPEI BRIGHT FRIGHT INC.
			Executed on (date) at (place) Signature of Issuing Carrier or its Agent
For Carrier's Use only at Destination	Charges at Destination	Total Collect Charges	BFI-12200

空運提單真的是份大單！許多的欄位與資訊就這樣塞進一份文件裡，它「排」「擠」得辛苦，我們可就更應該好好地、細細地來瞧一瞧，到底它英文說啥，中文是何意，這樣也才能讓那些只看少數幾個特定欄位的大多數人，終於有機會補足那一直以來的空白，一字一句地走一趟提單大地圖裡的每個角落。

一、左上區塊

297 8252 1000

Shipper's Name and Address 出貨人名稱與地址 拜爾肯科學公司 臺灣臺北市內湖區瑞光路337號 37樓	Shipper's Account Number 出貨人帳號
Consignee's Name and Address 收貨人名稱與地址 肯博公司 1000 XXXX 街 奇諾，加州 91710，美國	Consignee's Account Number 收貨人帳號
Issuing Carrier's Agent Name and City 開立提單之承運人代理名稱與城市 明亮貨運公司 臺灣臺北市信義區永吉路5號35樓	
Agent's IATA Code 代理人之國際航協代號 01-1-9000/0010	Account No. 帳號
Airport of Departure (Addr. of First Carrier) and Requested Routing 出發機場（啟運運送公司之地址）與要求路徑 TAOYUAN Airport 桃園機場	

To CHINO 至 奇諾	By First Carrier CI5100/10 透過啟運航空公司 CI5100/10	Routing and Destination 路徑與目的地	to 前往	By 透過	To 前往	By 透過

Airport of Destination 目的地機場 CHINO Airport 奇諾機場	Requested Flight/Date 要求之航班／日期	
	12/11/2022	

二、右上區塊

Not Negotiable 不具流通性 Air Waybill 空運提單 issued by 提單開立是由 BRight Freight Inc. 明亮貨運公司
Copies 1, 2 and 3 of this Air Waybill are originals and have the same validity 此空運提單第一、二、三份為正本聯，具同等效力。

It is agreed that the goods declared herein are accepted in apparent good order and condition (except as noted) for carriage SUBJECT TO THE CONDITIONS OF CONTRACT ON THE REVERSE HEREOF. ALL GOODS MAY BE CARRIED BY ANY OTHER MEANS INCLUDING ROAD OR ANY OTHER CARRIER UNLESS SPECIFIC CONTRARY INSTRUCTIONS ARE GIVEN HEREON BY THE SHIPPER, AND SHIPPER AGREES THAT THE SHIPMENT MAY BE CARRIED VIA INTERMEDIATE STOPPING PLACES WHICH THE CARRIER DEEMS APPROPRIATE. THE SHIPPER'S ATTENTION IS DRAWN TO THE NOTICE CONCERNING CARRIER'S LIMITATION OF LIABILITY. Shipper may increase such limitation of liability by declaring a higher value for carriage and paying a supplemental charge if required

茲同意在此所申報的貨品，依照背面運送契約條款所規定，送交的貨物皆具良好狀況除非另有說明。所有的貨物皆可透過包括公路運輸等方式運送，或由任何其他運送人運輸，除非託運人有不同的指示。此外，託運人同意貨物可運送至運送人認為合適的中途停留點。另請託運人注意，運送人的責任有限，如有需要，託運人可藉由對運送人申報較高的價值並支付附加費用，以擴充此責任限度。

Accounting Information	帳務資訊
Hoping Freight Ltd.	希望貨運有限公司
1500, XXXX AVE.	1500 XXXX 街
CHINO, CA 91710	奇諾，加州 91710
Tel電話: 909-548-6000	Fax傳真: 909-548-6111
Attn.: David	收件人：大衛

Reference Number 參考號碼	Optional Shipping Information 其他補充之出貨資訊	

Currency 幣別	CHGS 付款方式	WT/VAL航空運費／聲明價值附加費		Other 其他		Declared Value for Carriage 供運輸用所聲明之價值	Declared Value for Customs 供海關用所聲明之價值
US$ 美金		PPD 預付	COLL 預付	PPD 預付	COLL 預付		

Amount of Insurance 保險金額	INSURANCE - If carrier offers insurance, and such insurance is requested in accordance with the conditions thereof, indicate amount to be insured in figures in box marked "Amount of Insurance". 保險－若運送人有提供保險，而此保險為依其條件所要求，則請在箱子上註明保險金額為何，標註「保險金額」。
Handling Information 處理資訊	SCI 海關資訊

三、下半區塊

No. of Pieces RCP 件數 運價點	Gross Weight 毛重	kg lb. 公斤 磅	Rate Class 運價種類 Commodity Item No 貨品型號	Chargeable Weight 計費重量	Rate/ Charge 費率	Total 總計	Nature and Quantity of Goods (incl. Dimensions or Volume) 貨物之品名與數量（包含材積或體積）
2 CTNS 2個 紙板箱	50L 25K 50 公升 25 公斤		50L 25K 50公升 25公斤				Lab Research Reagents (Research Purpose Only) 實驗室研究用試劑（僅供研究用）

Prepaid 預付	Weight Charge 計費重量	Collect 到付	Other Charges Prepaid 其他費用 預付
Valuation Charge 估價費用			

<table>
<tr><td colspan="2">Tax 稅</td><td colspan="2" rowspan="6">I hereby certify that the particulars on the face hereof are correct and that insofar as any part of the consignment contains dangerous goods. I hereby certify that the contents of this consignment are fully and accurately described above by proper shipping name and are classified, packaged, marked and labeled, and in proper condition for carriage by air according to applicable national governmental regulations.

茲在此證明提單正面所記述內容正確，並在委託貨物含危險物品的規定範圍內。茲在此證明此委託貨物的內容已完整、正確地描述如上，記有其合適的出貨名稱、分類、包裝、標註與貼標，根據適用的國家政府規定，貨物亦適於空運運輸。

BIOCHEM SCIENCE 拜爾肯科學公司</td></tr>
<tr><td></td><td></td></tr>
<tr><td colspan="2">Total Other Charges Due Agent
代理人應付之其他費用總計</td></tr>
<tr><td></td><td></td></tr>
<tr><td colspan="2">Total Other Charges Due Carrier
運送人應付之其他費用總計</td></tr>
<tr><td></td><td></td></tr>
<tr><td></td><td></td><td colspan="2">Signature of Shipper or his Agent
託運人或其代理人簽名</td></tr>
<tr><td>Total Prepaid
預付總計</td><td>Total Collect
到付總計</td><td colspan="2" rowspan="3">AS AGENT FOR代理人
12/11/2014 TAIPEI BRIGHT FRIGHT INC.
臺北 明亮貨運公司</td></tr>
<tr><td></td><td></td></tr>
<tr><td>Currency Conversion Rate
幣別
匯率</td><td>CC Charges in Dest Currency
目的地到付費用</td></tr>
<tr><td></td><td></td><td colspan="2">Executed on (date) 執行日（日期） at (place) 執行地（地點）
Signature of Issuing Carrier or its Agent
製單運送人或其代理人之簽名</td></tr>
<tr><td>For Carrier' s Use only at Destination
僅供運送人於目的站時使用</td><td>Charges at Destination
目的站費用</td><td>Total Collect Charges
到付費用總計</td><td>BFI-12200</td></tr>
</table>

出貨文件–Commercial Invoice 商業發票 範例

Commercial Invoice

Invoice Number: SIS10501
Invoice Date: 12/01/22
Page: 1

Bill to: DRC International 1600, Schaefer AVE., CHINO, CA91710, USA		Ship to: DRC International 1600, Schaefer AVE., CHINO, CA91710, USA	
Sales Order Nbr: Customer P/O Nbr: Order Date: Salesman: Terms:	1041201005 SOC010747 12/01/22 Chris Chan Net 30	Ship Date: Method: Ship Via: Currency: AWB:	12/01/22 FOB Taipei FEDEX

Item No. Description	Ship Qty.	Size	Unit Price	Discount Pct	Total Price (USD)
AIG-00399 MAX Enzyme	5	100 units	500	10	2,250
Freight	1		50		50

Thank you for your order! All past due invoices are subject to a 1.5% monthly late charge. All prices include any applicable discounts and subvention obligations.	Subtotal: Discount: Total: Tax:	2,300 USD

出貨文件–Commercial Invoice 商業發票 範例　　中文

商業發票

發票號碼：SIS10501
發票日期：12/01/22
頁數：1

付款人：	收件人：
迪爾希國際公司 美國91710加州奇諾市薛佛路1600號	迪爾希國際公司 美國加州91710奇諾市薛佛路1600號

銷售單號碼：	1041201005		12/01/22
客戶訂貨單號碼：	SOC010747		FOB 臺北
訂單日期：	12/01/22		聯邦快遞
銷售人員：	詹克里斯		
付款條件：	淨30天		

型號 品名	出貨數量	尺寸大小	單價	折扣率	總金額(美金)
AIG-00399 麥克司酵素	5	100 顆	500	10	2,250
運費	1		50		50

謝謝您的訂單！ 所有逾期發票皆須收取1.5%每月遲付費用。 所有價格皆已含任何可適用之折扣及補貼金。	小計： 折扣： 總計： 稅：	2,300 USD

Show Time! 換你上場！

填空

No.	中文	英文
1	空運提單	
2	空運主提單	
3	空運分提單	
4	海運提單	
5	商業發票	
6	裝箱單	

選擇題

❶ 請問空運主提單單號的前三碼是：

(A) 阿拉伯數字 (B) 英文字母

❷ 請問空運分提單單號的前三碼是：

(A) 阿拉伯數字 (B) 英文字母

❸ 請問具有貨物所有權憑證之性質的是：

(A) 空運提單 (B) 海運提單

❹ 請問商業發票與裝箱單內容的差別在於：

(A) 型號 (B) 品名 (C) 數量 (D) 金額

翻譯

❶ 請見出貨明細如下，以及提單、裝箱單、商業發票如附。

❷ 依您所要求，在此附上一份完整的出貨文件，正本將會隨貨寄出。

來對對答案

填空

❶ Air Waybill

❷ Master Air Waybill

❸ House Air Waybill

❹ Bill of Lading

❺ Commercial Invoice

❻ Packing List 或 Packing Slip

選擇題

❶ (A)

❷ (B)

❸ (B)

❹ (D)

翻譯

❶ Please see below for your shipping details and the attached for the AWB, Packing Slip and Commercial Invoice.

❷ Per your request, please find the complete set of shipping documents attached. The original ones will be packed together with the shipment.

Part 2
內力修煉法

Part 2 篇章簡述

「People don't plan to fail; they just fail to plan.」，規劃很重要，企業要能永續經營，要維持利潤，進而強化企業的競爭優勢，一定得要好好地分析市場與競爭狀況，設定目標，擬訂整體行銷策略，設計並執行行銷活動，並且在執行後有完整的評估與檢討。

我們在Chapter 1中，會先來談談原廠吸引客戶多多訂貨的幾種促銷活動。促銷活動屬於短期的激勵措施，目的就在於要來好好刺激一下購買的意願與衝動！在許多種促銷、特惠方案中，最直白也最吸引人的促銷就是在價格上有折扣可享，例如買一送一的促銷方式，就可讓客戶明白地感受到花同樣的錢，居然可以享受得更多，有額外的好康。而這「買★送一」的條件與前提，可也有好幾種的不同內涵呢！例如買★的數量不同、買★的內容不同，以及買★送☆的設計搭配。這些「如懿傳」的不同風貌，在1-1中會有好些著墨，再請你慢慢看看能怎麼地如意賺。

接著在1-2裡，我們就要一塊兒來瞧瞧另一種大熱門的促銷方法：「滿額折扣」。這種促銷的好康活動，也有好幾種不同的設計考量方向，從折扣的類型（例如所折為一件產品或固定金額）、折扣的對應標的（例如針對某特定產品、特選產品）、折扣兌換的類型（例如折扣碼或折扣券），到折扣兌換的時間，每一個考量方向都可設計出許多不同的好康與條件方案，可讓客戶有那種付了一定的錢，卻能賺多多的感覺。

而這些對於促銷特惠活動的規劃，不管促銷規模大小，一定都要納入適用與不適用的限制條件。在**1-3**中，我們就藉助基本又好用的**5W1H**法則，來逐一看看有哪幾種限制條件是我們在規劃促銷活動時必不可馬虎、必得反覆「佈局」的地方！

到了Chapter 2，我們可就要來看看「參展」這另一種行銷模式了！參展是「面對面行銷」，它可是有著其他行銷工具所無法取代的特性。在**2-1**中，我們會針對原廠自己參展的部分來探討，看看原廠要如何通知客戶、如何邀請客戶來參觀展覽，共襄盛舉。另外，國際上舉行那麼多種的會議與展覽，有的叫Exhibition，有的稱為Conference，有的就叫做Meeting……，到底可稱為會議或展覽的字詞有哪些呢？這些不同字詞彼此之間是不同在哪裡呢？在**2-1**的「關鍵字急救站」裡，我們就要把這些會展的字詞一次說個明白！而除了這些關鍵字之外，我們還會細細說明參展的目的，看看原廠想藉著參展在在銷售方面達到什麼樣的目標。此外，原廠除了自己報名參展之外，當各地經銷商參展時，原廠一樣也要提供最大的支援，以能一同創造更多的業績，齊力提升全球的市佔率……這就是我們在**2-2**所要說明的主題，而除了說明支援上的溝通情境之外，我們也會來瞧瞧原廠支援經銷商參展的幾種不同的方式。

行銷活動安排了，展覽也參加了，那到底這些在行銷上的努力有沒有反應在企業的實質業績上呢？業績的表現如何？跟以前的實績比起來是增加還是減少呢？這些管控與檢討工作，可是重要得不得了！有了管控、有對成果進行檢討與評估，才能知道企業的績效表現，才能知道企業在市場競爭上的位置，也才有辦法即時調整行銷的方向與策略！在Chapter 3這最後一章中，我們就要來一起看看原廠與經銷商針對業績檢討與年度檢討報告的溝通內容與內涵！

多多訂貨

1-1 誘人如懿傳 — 買★送一如意賺

Cathy: Hi. This is Cathy Hsu. Is Roger there, please?

Roger: Roger speaking. How are you, Cathy?

Cathy: I'm very well. And you?

Roger: I'm good! Thanks.

Cathy: I'm calling to tell you good news! We're planning to launch our buy 2 get 1 free promotion!

Roger: How great it is! Does the promotion cover all your products?

Cathy: It applies to all our antibody products. Customers can receive a secondary antibody for free when they order 2 primary antibodies within the same order.

Roger: That will meet the best interests of our customers! When will the promotion start and end?

Cathy: We provide an extensive promotion duration from April 1 to September 30 to allow your customers to make the best use of it!

Roger: I'm sure we'll get great sales results!

Cathy: That's exactly the answer I would expect from you!

背景說明	
人物	Cathy：原廠業務經理 Roger：經銷商產品經理
主題	原廠推出買二送一的促銷活動： 促銷產品類別、贈送產品類別 & 促銷期間

譯文

凱西： 嗨，我是許凱西，請問羅傑在嗎？

羅傑： 我就是，妳好嗎，凱西？

凱西： 我很好，你呢？

羅傑： 也不錯，謝謝。

凱西： 我打來是要告訴你一個好消息！我們準備要推出買二送一的促銷活動呢！

羅傑： 很好耶！那這個促銷活動包含妳們所有的產品嗎？

凱西： 這個活動適用我們所有的抗體產品，只要客戶一次下單訂購兩支一級抗體，就可得到一支免費的二級抗體。

羅傑： 這最合我們客戶的興趣呢！請問促銷活動什麼時候開始？到何時結束呢？

凱西： 我們設定了一個頗長的促銷期，從4月1日開始，到9月30日結束，好讓你的客戶能夠好好利用呢！

羅傑： 我相信我們會做出很好的業績！

凱西： 這就是我希望從你這兒聽到的回答呢！

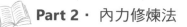

說三道四，換句話試試

例一：好康報報標題

例句一　Special Promotion: Buy 1 Get 1 Free

譯文一　特別促銷：買一送一

例句二　Buy 2 Get 1 Free on Selected Items

譯文二　精選品項，買二送一

例句三　Limited Time Offer – Buy 3 Get a 4th Free

譯文三　限時特惠 — 買滿三個，第四個免費

例句四　Buy 4 Get 1 Free of the same item

譯文四　買四就送一個相同產品

例句五　Take advantage of our 2023 Springtime special offer!

譯文五　請多加利用我們的2023春日特惠！

例二：好康期間

例句一　Offer valid until July 31

譯文一　特惠期間至7月31日為止

例句二　Offer ends on July 31

譯文二　特惠將於7月31日結束

例句三　Two Weeks Only!

譯文三　特惠期間只有兩個星期！

例句四　The offer is valid for orders until end of July, 2023

譯文四　2023年7月底之前下單都適用此特惠

例句五　Act quickly! You only have until the December 24 to place your order to be eligible for this fantastic offer!

譯文五　快快行動！12月24日之前下單才能享有這個超值特惠！

例三：如何利用好康？要寫促銷編碼

例句一 Please use promotion code: PYC1202.

譯文一 請註明促銷編碼：PYC1202。

例句二 Simply enter this code at the checkout: PYC1202FREE or quote it when you order by email, fax or phone.

譯文二 只需在結帳時輸入PYC1202FREE，或是在Email、傳真或打電話下單時註明此編碼。

例句三 Please include promotion code "FREE2ND" for all three items and list the final price for free secondary antibodies as $0. A sale order example is attached in this Email for your reference.

譯文三 三個品項都請加註促銷編碼「FREE2ND」，在免費的二級抗體金額處請列＄0。在此Email中，我們另附上一份銷售訂單範本，供您參考。

關鍵字急救站

我們在行銷這部份要談的第一個關鍵字就是「promotion」／「促銷」！讓我們一起看看promotion的字首、字根、字尾，以及它的搭配詞、句子與使用情境，讓你跟promotion這個字從原來的初識，進展到火熱的熟識程度，讓你一次就徹底地把它摸透透，之後在寫Email或是説電話時，就能下「指」快、開口溜，述說句子有信心！對於本書中所列的每個關鍵字，請跟著所寫的一字一句慢慢看下去、慢慢修煉，不要輕易放過瞭解字詞的機會，也不要隨便看待每一個字！只要你肯讓字詞在腦中多待一會兒，多想個一回，那你對該字詞就能參透得多一些，而當你慢慢積累這樣的「一些些」之後，你懂的、你熟的就不只有「一些些」了啊！

讓我們回到這一個單元的關鍵字。首先，我們來拆解一下promotion：

- 字首 pro：意思為 for、forward／贊成、向前
- 字根 mot：意思為 move／移動
- 字尾 tion：為常見的名詞字尾

所以動詞promote就是「向前移動」，是「to support or encourage something」／「支持或鼓勵某事」的意思，會promote，就會「contribute to progress or growth」／「促成進步或成長」。而名詞promotion就是「the process of attracting people's attention to a product or event」，是促銷，是讓消費者受到某一項產品或活動吸引的過程。

接著，我們就來看看promote、promotion與promotional在各種情境下的使用實例囉：

一、動詞 promote

例 We're planning to promote and supply your product to the customer.

我們打算促銷您的產品，提供給客戶。

例 The Distributor shall offer technical support to the customers and use its best efforts to promote the sale and distribution of our products.

經銷商需要向客戶提供技術支援，盡力推廣和分銷我們的產品。

例 By this letter, we state that Goal Biomedical, Inc. with address at: 1202 Wall Street, New York, NY 10286, USA, is an authorized representative of Eden Technology in the Territory of USA, to sell and promote all products made by our company.

茲以此信函聲明高爾生醫公司，設址於美國紐約10286華爾街1202號，為伊甸科技在美國地區所授權的代理商，以銷售與推廣我們公司生產的所有產品。

二、名詞 promotion

例 The promotion is valid for both domestic and international customers.

國內與海外客戶都適用此促銷方案。

例 We will keep you informed of new products coming online along with the products that are currently on promotion.

若線上有推出新產品或是有新的促銷產品，我們就會通知您。

三、形容詞 promotional

例 We will be able to send 2 boxes of bags along with the other promotional materials to you no later than October 2.

我們最晚會在10月2日之前，將兩箱的袋子連同其他促銷用品寄給您。

例 We would be happy to work with you to provide some promotional items to distribute to visitors.

我們很樂意配合提供一些促銷禮品，讓您發送給參觀的人。

例 Also please share with me any promotional campaigns that you

plan to run throughout this year, so we could know how to best help you.
請告訴我你們今年一整年所計畫的促銷活動，這樣我們也好盡力來協助你們。

術語直達車，專業補給站

「如懿傳」的不同風貌！

買一送一的這類銷售促銷方案，可給客戶那種付了一定的錢，卻能賺多多的感覺。 “Sales promotions such as buy-one-get-one-free (BOGOF) give customers more for their money!”

你一定聽過或看過「如懿」，但你有沒有每次都「如懿」呢？很有可能沒有！因為她有她的性情與脾氣，但你有你的想像和期許,「如懿」很美，但不必然是為了你而美麗！這...這...這是說到哪兒去了啊？…「如懿」就是如意賺，「如意賺」有寬廣的想像空間，也有不見得盡如人意的實情與全貌，請從下文見分曉囉！

一、買★的數量不同

這個好理解，不同的數量就像是有買一送一、買二送一、買三送一、買四送一……我想我要是繼續列下去了，除了明顯呼嚨讀者之外，也有促銷活動落入愈來愈不吸引人的窘境了啊……你想想，買一百送一？！這也太沒有促銷的誠意了吧！不過，若是以犒賞忠誠客戶、維繫合作關係，就也還適合設計買到五個以上就送一個的方式，以獎勵這些長期客戶的忠心……「A program of Buy Five and Get the Sixth Free can help retain customers and boost sales.」，就是鼓勵客戶買到五個，第六個就可免費贈送，這樣的的促銷方案就是設計用來幫助留住客戶，也增加銷量。

二、買★送★的內容不同

舉「買三」的促銷設計來説，送一的前提條件可以有這些不同的方式：

- 相同產品：同樣的產品買個三件的話，就可享有送一個相同產品的優惠，例如說道：「Buy 3 Get 1 Free of the same catalog number」。
- 沒設限的不同產品：所有產品中任選個三件，就能送一，例如乾乾脆脆的「Buy 3 Get 1 Free」，後頭沒跟任何但書與限制，或是例如「4 for 3 offer. Order any three products at list price and receive a fourth product of equal or lesser value FREE!」，產品不限，第四個同價或價較低的產品免費。
- 有限制的不同產品：所買的三個產品可為不同品項，但須為同一類的產品，例如：
  - Buy any three R-series Kits and get a forth R-series Kit of equal or lesser value for free.
    買任三組R系列套組產品，就可享有同價或價較低之第四組R系列套組的免費優惠。
  - Buy 3 Get 1 Free on Gloves!
    手套特惠，買三送一！

三、買★送☆

買牛肉是否就送牛肉呢？喔不！我說送一，可沒說就要送牛肉，也可以送豬肉、雞肉或羊肉呢！因為「送一」條件會強烈吸引消費者的目光，當消費者一被引了過來，就有成交的可能，所以，在有些促銷活動中，客戶所買的標的產品，跟廠商所送的產品並不相同。通常所送產品的單價會低於所買產品的單價，像是潛水用品商家就有可能推出二級頭套組特惠：「Buy 1 Get 1 Free on Second Stages」來吸引消費者，買的一與送的一，一樣都是二級頭，不過套組裡一個是質優的二級頭，送的可能就是一個較為便宜的「章魚式」備用二級頭了。

廠商可以精心設計這些「送一」行銷方式，以達到吸睛、提振業績的功效，但請注意，所有「設計」的內容與細節，都要能在促銷通知中讓客戶一目瞭然，設計的本質在於吸睛，不是讓客戶有在糊里糊塗中受騙上當的感覺！

此外，每當我們看到買一送一這類促銷方式的分析時，常也同時會看到這樣的提醒 — 買一送一會引來客戶搶購，但其實它也是有風險的……它會增加銷量，但可能會削減利潤，還有更重要的一件事，這樣的促銷活動若常常推出，那可能會削弱品牌形象與價值！接下來，就讓我們來看看這些風險的中英文說明囉：

➡ 削減利潤……

If customers have bought two for the price of one, unless sales promotions attract new customers, the overall effect may be to reduce profits. So while sales promotion can work well for new product launches, there are more risks with existing products.

若是客戶以買「一」的價格得到了「二」，除非銷售促銷的目的是在於吸引新客戶，否則整體的結果可能是削減了利潤。因此，當銷售促銷方案對新產品上市的推廣有助益時，對於原有產品可能是帶來了更多的風險。

➡ 削弱品牌形象與價值

We should also think carefully about how sales promotions affect our brand. Regular price discounting might devalue our brand.

我們應該仔細思考究竟銷售促銷方案對我們的品牌形象會造成什麼樣的影響，常態的價格折扣反而也有可能削弱了我們的品牌價值。

這些是從另一個角度來評估「送一」這類行銷手法的影響，我們在設計行銷活動時，亦應知曉這樣的風險，以能在設計行銷方案時時好好地一併考量，做出得宜且最適的活動設計！

最後，我們就來看看廠商發給經銷商的促銷方案通知實例囉！

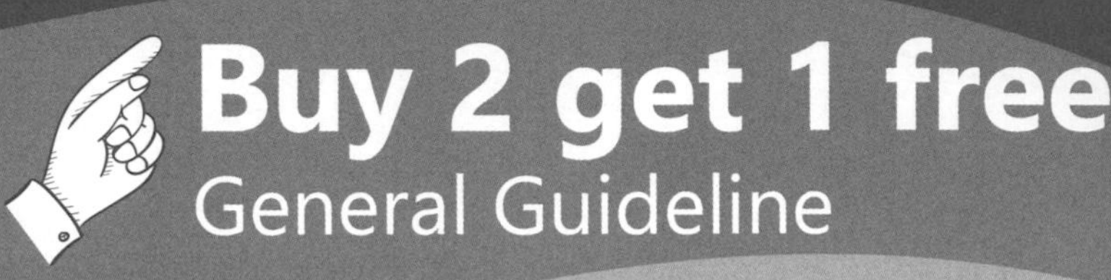

Dear valued partners,

We are happy to inform you that Summit Biomed is offering a promotion to our distributors from March 1 to August 31 to help promote Summit products in your territory. Here are some general guidelines for this promotion.

- Customers can receive a secondary antibody for free when they order 2 primary antibodies from Summit (within the same order).

- Promotion Duration: March 1 to August 31, 2023.

- Please include promotion code "2NDFREE" for all three items and list the final price for free secondary antibodies as $0. A sale order example is attached in this email.

- Order and logistic process remain the same unless otherwise notified.

- All secondary antibodies are eligible for this promotion. (Summit's complete secondary product list is attached.)

- We also have a list of 10 popular secondary antibodies as our featured giveaway. Summit will have them in stock during promotion period.

- Summit logo is attached for your marketing purpose.

Please kindly reply to this notification if you would like to join this promotion. Feedback or questions are also welcome.

Thank you for promoting Summit!

Sincerely,

Hank Chen
Marketing Manager
Summit Biomed

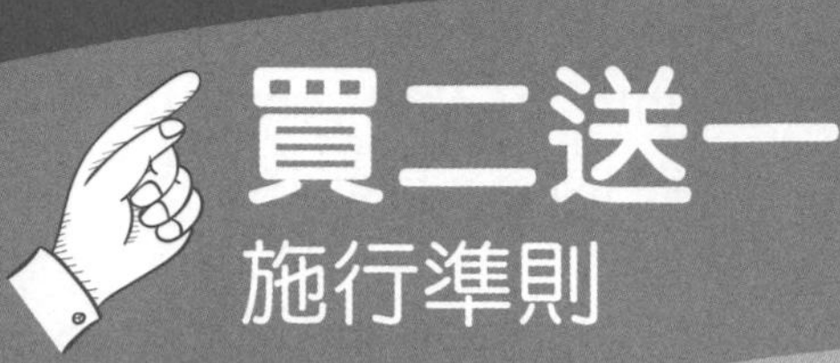

買二送一
施行準則

各位重要的夥伴，您們好，

我們很高興地要在此通知您，高峰生醫將要推出一項促銷方案給我們的經銷商，期間將自3月1日開始，至8月31日日結束，以協助您們在當地行銷高峰的產品。下列為此促銷方案的幾項施行準則：

- 客戶若訂購兩支一級抗體，則可得到一支免費的二級抗體（同一張訂單內訂購）。

- 促銷期間：2023年3月1日至8月31日

- 下單訂購這三項產品時，請加註促銷編碼「2NDFREE」，而所贈送的免費二級抗體，價格請列為 $0。另附上銷售確認單範例，供您們參考。

- 下單與作業程序將維持不變，除非另有通知。

- 所有的二級抗體皆在此促銷案範圍之內。（高峰的二級抗體完整產品表如附）

➢ 我們亦列出了十項熱銷的二級抗體為此促銷之主力贈品。高峰會在促銷期間維持其現貨供應狀況。

➢ 附上高峰的商標圖案，供您們行銷所需。

若您們想要加入此促銷方案之列，請給我們個回覆，歡迎您回饋意見，若有任何問題，也歡迎提問。

謝謝您們為高峰促銷所做的努力！

謹上

陳漢克
行銷經理
高峰生醫

Show Time! 換你上場！

促銷標題翻譯

❶ 買一送一

____

❷ 買滿三個，第四個免費

____

❸ 精選品項，買二送一

____

❹ 相同產品買四就送一

____

好康期間翻譯

❶ 限時特惠

____

❷ 特惠效期2023年2月28日為止

____

❸ 特惠將於10月30日結束

____

❹ 2023年7月底之前下單都適用此特惠。

____

❺ 快快行動！12月24日之前下單才能享有這個超值特惠！

____

來對對答案

促銷標題翻譯

❶ Buy 1 Get 1 Free

❷ Buy 3 Get a 4th Free

❸ Buy 2 Get 1 Free on Selected Items

❹ Buy 4 Get 1 Free of the same catalog number（或是the same product）

好康期間翻譯

❶ Limited Time Offer

❷ Offer valid until February 28, 2023

❸ Offer ends on October 30

❹ The offer is valid for orders until end of July, 2023.

❺ Act quickly! You only have until the December 24 to place your order to be eligible for this fantastic offer!

Chapter 1 多多訂貨

1-2 滿額折扣

Albert: Hello, this is Albert from Pioneer AutoTech. May I speak to Susan Dawson?

Susan: Hi, Albert. This is Susan. How are you today?

Albert: Not bad! I have news for you. We want to start a new campaign for the next 6 months!

Susan: It's good news! What are your plans for the campaign?

Albert: Any customer who buys from our variety of products at a total of more than 2,000 USD will get 25% discount on the next order.

Susan: So, when we place our order for you for various customers, we need to specify the name of the customers who exploit this campaign, right?

Albert: Correct. Also please note that, between these two orders, there should be an interval of at least 1 month and no longer than 6 months.

Susan: I see. When will the campaign start?

Albert: It will be published at our Newsletter next week.

Susan: That's fine. I could share the information with our key customers first and encourage them to take advantage of this special offer!

Albert: Great! Thanks!

背景說明	
人物	Albert：原廠行銷經理 Susan：經銷商業務經理
主題	促銷方案通知： 滿2,000美元，享25%off，下個訂單折抵！

亞伯特：哈囉，我是先驅自動科技的亞伯特，麻煩請找蘇珊．道森。

蘇　珊：嗨，亞伯特，我就是，你今天好嗎？

亞伯特：還不錯呢！我有個消息要告訴妳，我們準備要在接下來的六個月推出一個新的促銷活動！

蘇　珊：這是個好消息！這個促銷活動是怎麼樣的呢？

亞伯特：客戶購買我們的各種產品，若是總金額超過2,000美元，就可享有25%的折扣，下次訂單就可扣抵。

蘇　珊：所以我們一次幫多個客戶下單時，就得寫明要用此促銷活動的客戶名，是嗎？

亞伯特：沒錯，也請注意，兩次下單間隔的時間至少要一個月，但也不能超過六個月。

蘇　珊：瞭解，那這促銷活動什麼時候開始呢？

亞伯特：我們下星期的電子報就會公布這個活動的消息。

蘇　珊：好的，我也可以先跟我們的重要客戶說說這個消息，也請他們努力多多利用這個特價優惠！

亞伯特：太棒了！謝謝！

說三道四，換句話試試

滿額折扣好康實例

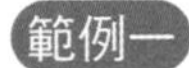

範例一

ORDER MORE, SAVE MORE

No Discount Code Needed! Your savings will be automatically deducted from your order.

Order More Than	Save
$200	$20 off your order
$400	$60 off your order
$600	$120 off your order
$800	$200 off your order

譯文一

訂愈多，省愈多

不需要用折扣碼！省下來的金額會自動從您的訂單中扣除。

訂單金額超過	省多多
$200	訂單直接扣 $20
$400	訂單直接扣 $60
$600	訂單直接扣 $120
$800	訂單直接扣 $200

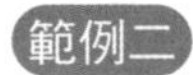

範例二

SPECIAL OFFER

This summer you will be thrilled by our sensational discounts!

Order amount: More than	Save
US$1,000	10%
US$1,500	15%
US$2,000	20%
US$2,500	25%
US$3,000	30%

譯文二

特惠促銷

讓您興奮一「夏」的感人折扣！

訂單金額：超過	折扣
1,000美元	10%
1,500美元	15%
2,000美元	20%
2,500美元	25%
3,000美元	30%

範例三

GET A DISCOUNT ON YOUR ORDER

The higher the order amount, the higher the discount!

BUY FOR $500 AND GET 5% DISCOUNT

BY USING COUPON CODE: DISC5

BUY FOR $1,000 AND GET 10% DISCOUNT

BY USING COUPON CODE: DISC10

BUY FOR $1,500 AND GET 15% DISCOUNT

BY USING COUPON CODE: DISC15

譯文三

下單賺折扣
訂單金額愈高，可享折扣愈高！
買滿 $500，就有 5% 的折扣 請使用折扣券編碼：DISC5 買滿 $1,000，就有 10%的折扣 請使用折扣券編碼：DISC10 買滿 $1,500，就有 15%的折扣 請使用折扣券編碼：DISC15

範例四

With larger orders, exceptional value can be achieved!

Our discounts are as follows:

Order Amount	Save
US$500+	10%
US$1,500+	20%
US$3,000+	30%
US$5,000+	40%

譯文四

您給我大訂單，我給您非比尋常的驚喜價！

折扣大放送如下：

訂單金額	折扣
500美元	10%
1,500美元	20%
3,000美元	30%
5,000美元	40%

範例五

Summer Kick-off Sale!

15% OFF!

HURRY! DON'T MISS YOUR CHANCE!

Get 15% OFF if you order more than US$500!

Select from thousands of high quality products!

Offer valid from 1/1/23 to 6/30/23

Promo code: BXSUMMER16

譯文五

夏日熱銷開跑！

85 折！

快快快！不要錯失大好機會！

只要訂滿 500美元，就可享 85 折的好康折扣！

數千項高品質的產品任你選購！

好康期間：1/1/23～6/30/23

促銷碼：BXSUMMER16

關鍵字急救站

在說到促銷方案與活動時，常會用到這個字：「campaign」，我們就藉此機會來好好對它品頭論足一番！

➡ campaign (n.) [kæm`pen] 活動

英英解釋：a series of things such as television advertisements or posters that try to persuade people to

buy a product
用來說服客戶購買產品的一系列活動，像是電視廣告或海報

Campaign這個字源自於拉丁文字「campania」，意思為「open field; plain; level countryside」，亦即平坦的田野，之後此拉丁文也有了「戰場」的意思，而campaign就演變成了軍事上所說的「戰役」，後來也有了「競選活動」的意思，以及商業上所指的「促銷活動」。

Campaign此字指促銷活動時，常見的搭配詞為marketing campaign／行銷活動，現在就讓我們來看看它的情境與用法囉！

情境一

原廠請經銷商告知計劃推出的促銷活動有哪些，以能瞭解一下經銷商對其品牌推廣的準備，亦可給予些建議。

例 Please share with me any promotional campaigns that you plan to run throughout this year, so we could know how to best help you.
請告訴我你們今年一整年所計劃的促銷活動，這樣我們也好盡力來協助你們。

情境二

原廠告知經銷商其行銷部門會提供有關行銷活動方面的協助。

例 Our Marketing Dept. will help you plan and answer all your questions about marketing campaigns.
我們的行銷部門會幫助您規劃，並回答您所有有關行銷活動的問題。

情境三

原廠已在美國推出了一個行銷活動，現提供相關明細與資料給其他區域的經銷商，讓他們也可在其本地市場推行同樣的促銷活動。

例 We launched a marketing campaign in the US market. If you'd like to mirror our campaign in your local market, you may find its details in the attachment for your reference.

我們在美國市場有推出了一個行銷活動，若是您想要在您本地市場仿效推行，我們在此提供活動的明細如附，供您參考。

術語直達車，專業補給站

我們在上一節看了「送一」，這一節瞧了「滿額折扣」，究竟這類行銷、促銷好康活動大致上可分出哪幾種不同的方式呢？讓我們深呼吸一下，來一口氣地好好「數落」一番吧！

➡ 折扣的類型：When meeting a minimum of order quantity or amount, discounts are given as:

- a product／一件產品
- a percentage／一個百分比
- a fixed amount／固定金額
- free shipping／免運費

折扣的類型可以是產品，例如免費多送一件，以及這一節所說的滿額折扣。而所折抵金額的算法，就有前例中所提及的「Order more than US$1,000, discount up to 10%」這種折抵固定比例的算法，以及「Order more than US$200, save US$20 off your order」以固定金額來折抵的方式，當然還有設計上最利便的免運費好康囉！

➡ 折扣的對應標的：Applying a discount to

■ a certain product／某特定產品

■ selected products／精選產品

■ the total amount of an order／訂單總金額

■ a certain customer／某特定客戶

若要針對某項或某類產品來做促銷，例如每月強打品項或是新上市的產品，則折扣條件就會針對該項、該類產品來設計。若是要吸引客戶多訂，衝一下廠商的銷售額，則所推的折扣就會是從整單、總金額來扣抵。此外，也有重要客戶、專案，須以特殊價格與折扣來因應激烈競爭的需求。

➡ 折扣兌換的類型：Applying a discount by using

■ discount codes／折扣碼

■ coupons／折扣券

出口廠商要分辨不同行銷活動所給予的不同折扣內容，會給予不同折扣碼與折扣券來讓客戶下單時註明，以能正確抵扣。例如前例中「BUY FOR US$500 AND GET 5% DISCOUNT, by using COUPON CODE: Disc5」，就是以不同的折扣碼來分辨不同的滿額折扣率。而在廠商參展時，就常會發送該展覽的促銷折扣券來吸引客戶前往參觀，吸引客戶在展覽中下單。

➡ 折扣兌換的時間：Applying a discount

■ at the time of ordering／該筆訂單

■ on next order／下一筆訂單

■on next order within a certain period of time／期限前的下一筆訂單

出口廠商所設計的折扣兌換時間，可為該筆訂單下單時，例如買三送一就會是該訂單出貨時，一併出貨所送的產品，或是下次下單時，吸引客戶再來下單一次。而折扣所能適用的下一個訂單的下單時間，常也會規定要在一個期間內，例如前面「對話」中所說到的「there should be an interval of at least 1 month and not longer than 6 months」，兩個訂單不能隔太近，以防有人刻意將原要下的一個訂單拆成兩次分開下單，但也不能隔太遠，以能吸引並提醒客戶要早點下單呢！

促銷標題翻譯

❶ 訂愈多，省愈多！

____

❷ 訂單金額愈高，可享折扣愈高！

____

❸ 訂單超過1,000美元，折扣可達10%！

____

❹ 訂單金額超過200美元，訂單可省20美元！

____

❺ 夏日熱銷開跑！

____

促銷事項翻譯

❶ 請告訴我你們今年一整年所計畫的促銷活動，這樣我們也好盡力來協助你們。

❷ 不需要用折扣碼！省下來的金額會自動從您的訂單中扣除。

❸ 我們在美國市場有推出了一個行銷活動，您可在您本地的市場仿效推行。

❹ 任何客戶購買了我們多種產品的總金額若超過2,000美元，他的下個訂單就可享有25%的折扣。

促銷標題翻譯

❶ Order more, save more!

❷ The higher the order amount, the higher the discount!

❸ Order more than US$1,000, discount up to 10%!

❹ Order more than US$200, save US$20 off your order!

❺ Summer Kick-off Sale!

促銷事項翻譯

❶ Please share with me any promotional campaigns that you plan to run throughout this year, so we could know how to best help you.

❷ No Discount Code Needed! Your savings will be automatically deducted from your order.

❸ We launched a marketing campaign in the US market. You could mirror our campaign in your local market.

❹ Any customer who buys from our variety of products at a total of more than US$2,000, he'll get 25% discount on his next order.

多多訂貨

1-3 特惠的限制條件

Lucy: Hello, Carl. This is Lucy from Delta Science.

Carl: Hi, Lucy. What can I do for you?

Lucy: I noticed on your website that you have a marketing campaign of Summer Sale 20% Off. Is this campaign also available to distributors?

Carl: I'm afraid I have to let you down… On our website, you could also find some restrictions to the promotion. One of them says that this discount cannot be combined with any other offers, promotions, or discounts offered by us.

Lucy: I didn't notice that. So this promotion applies to end users only, right?

Carl: Yes, it's open to end users placing orders online. Not eligible for orders received by Email or fax or through distributors.

Lucy: I see. So when could we expect to have a promotion for distributors like us?

Carl: Hahhah... Very soon! Aren't you going to attend an exhibition next Month? We're working on a punchy promotion for you to attract visitors to place orders at the exhibition!

Lucy: That's great! Thanks in advance for your support! I'll wait for your good news!

背景說明	
人物	Lucy：經銷商產品經理 Carl：原廠行銷主任
主題	經銷商確認可否適用促銷方案： 促銷方案僅供網路下單的客戶使用，下個月經銷商參展會支援並提出適用的折扣方案。

譯文

露西：哈囉，卡爾，我是三角洲科技公司的露西。

卡爾：嗨，露西，有什麼需要我幫忙的嗎？

露西：我從你們的網站上看到「夏日促銷，現折20%」的行銷活動，請問經銷商也可適用這個活動嗎？

卡爾：恐怕我得讓妳失望了……在我們的網站上也有列出這個促銷活動的一些限制，其中一條有說到這個折扣不能與我們所提供的其他任何報價、促銷或折扣併用。

露西：我沒注意到，所以這個折扣只適用最終使用者，對嗎？

卡爾：是的，這折扣適用網路下單的最終使用者，用Email、傳真，或透過經銷商所下的訂單，都不適用。

露西：瞭解，那何時會有給我們這些經銷商的促銷方案呢？

卡爾：哈哈……快了快了！你們不是下個月要參展嗎？我們現在正在研擬一個強而有力的促銷方案，讓你們能在展覽上吸引參觀的人下單呢！

露西：太棒了！先謝謝你們的支援！那我就等你的好消息囉！

說三道四，換句話試試

例一 | 限對象 & 區域

例句一 Open to end users of the United States only.

譯文一 僅適用美國區域的最終使用者。

例句二 Valid in US & Canada only.

譯文二 僅適用美國與加拿大區域。

例句三 Offer valid for European customers only.

譯文三 特惠僅適用於歐洲客戶。

例二 | 限時

例句一 Offer expires on December 31, 2023.

譯文一 特惠至2023年12月31日為止。

例句二 All orders received online by March 2, 2023 with the discount code will be eligible for the promotion.

譯文二 在2023年3月2日前有標註折扣碼的所有網路訂單，皆可適用此促銷方案。

例句三 Offer will apply to qualifying orders received by Top Biosciences between July 1–August 31, 2023.

譯文三 頂尖生科在2023年7月1日至8月31日期間所收到的所有合格訂單，皆可適用此特惠。

例三 | 限次

例句一 One time use only.

譯文一 僅限使用一次。

例句二 Limit one free item per customer.

譯文二 每個客戶僅限兌換一次免費品項。

例句三 Apply to a single purchase.

譯文三 僅限一次訂購使用。

例句四 Customers can use the discount no more than once during the promotion period.

譯文四 此折扣僅供客戶在促銷期間使用一次。

例四 | 折扣不得併用

例句一 Discount cannot be combined with any other promotions or discounts offered by Top Biosciences.

譯文一 折扣不能與頂尖生科其他的促銷方案或折扣併用。

例句二 Cannot be combined with institutional discounts, other discounts or promotions.

譯文二 不能與機構折扣、其他折扣或促銷方案併用。

例句三 Limit 1 promotion per order.

譯文三 每張訂單僅適用一種促銷方案。

例句四 You can take advantage of all our 6 special offers, but offers cannot be combined on the same order.

譯文四 您可利用我們的六項特惠方案，但不得併用在同一張訂單裡。

例五 | 限新訂單

例句一 Does not apply to past purchases.

譯文一 不適用於先前的訂單。

例句二 Offer may not be applied to existing, pending or prior orders.

譯文二 特惠不適用於已下單、尚未出貨或先前的訂單。

例句三 Offer valid on new orders only.

譯文三 特惠僅適用於新訂單。

例六｜關於產品品項的限制

例句一 Does not apply to A series products.

譯文一 不適用於A系列產品。

例句二 Free product is Cat #D1202. No substitutions or exchanges for free product.

譯文二 所提供之免費產品為型號D1202，不得改換其他品項。

例句三 Limited to Research Use Only products.

譯文三 限「僅供研究用」之產品。

例句四 Free item to be of equal or lesser value.

譯文四 免費品項須為等值或價值較低的品項。

例句五 Free item must be of equal or lesser value to the lowest priced purchased item.

譯文五 免費品項須為與所購買金額最低之產品等值或價值較低的品項。

例句六 Offer excludes custom orders, services, and kit products.

譯文六 特惠不適用於訂製訂單、服務，以及套組產品。

例句七 Offer applies to purchases of 3 catalog items or more and excludes 100ml or smaller sized products.

譯文七 特惠僅適用於訂購三個或三個以上品項之訂單，不包含100毫升或低於100毫升之規格的產品。

例七 | 要標註促銷碼

例句一 Promotion code must be presented at the time of order.

譯文一 下單時須加註促銷碼。

例句二 Promotion code OMG15 must be referenced when placing order.

譯文二 下單時須加註OMG15促銷碼。

例八 | 原廠所保留之權利

例句一 Top Biosciences reserves the rights to modify or cancel an offer at any time.

譯文一 頂尖生科保留在任何時間修改或取消特惠的權利。

例句二 Other restrictions may apply.

譯文二 亦可能有其他的限制條件。

例九 | 其他限制與注意條件

例句一 No substitutions.

譯文一 折扣不得替換。

例句二 No cash or cash equivalent.

譯文二 不得兌現或以現金等價物兌換。

例句三 Items that are not eligible, tax and shipping and handling charges will not be included in determining merchandise subtotal.

譯文三 不合乎條件之品項、稅金，以及出貨與處理費，將不計入產品小計之中。

例句四 Credit card order confirmations will not reflect the discount, as the discount will be applied at the time of invoicing.

譯文四 信用卡訂貨確認單上尚不會反映出折扣，因折扣會是在開立發票時扣除使用。

例句五 Offer void where prohibited, licensed, or restricted by any laws or regulation or institutional policy.

譯文五 凡屬任何法律、法規或機構政策所禁止、以許可規範或限制者，則此特惠將屬無效。

關鍵字急救站—restriction

在促銷特惠活動的通知訊息裡，除了一定會有吸引人的好康價格、折扣及贈品內容之外，一定也要將限制條件一併說明清楚，才不會讓原廠忙著事後個別解釋與說明，也不會讓方案上制定為不適用的客戶空歡喜一場！所以，我們在這兒就來看看所謂的「限制」／restriction囉！

➡ restriction (n.) [rɪˋstrɪkʃən] 限制

英英解釋：a rule, action, or situation that limits or controls someone or something

限制或控制某人、某事的規則、行動或情況

restriction的字形可拆解如下：

字首「re-」／重複

＋ 字根「strict」／嚴格的

＋ 字尾「-tion」／名詞詞性

從意義上來理解，當一件事一直「重複」地執行某項「嚴格的」管控時，這事項就成了此事的「限制」條件了。在國貿上常見的情境與用法如下：

情境一

屬於限制進口或出口的產品，國家會實行配額或者許可證管理

例 Some products may be restricted in certain countries.

有些產品可能在某些國家有限制管制。

例 Unfortunately, at this time, we are not able to sell our products to your country, due to licensing restrictions.

抱歉目前因為有授權許可的限制，所以我們無法銷售產品到您的國家。

例 Product #0925 has no export restrictions at all. None of the components are considered controlled by our local authority.

型號0925的產品並沒有任何的出口限制，其所含成份沒有任何一項是在我們本地主管機關管制的範圍之內。

例 Please try to see if you can purchase these products on your end, and we will check and let you know of any our export restrictions.

請試著查查是否您可否購買這些產品，我們也會查查有無任何的出口限制。

情境二

有設價格限制

例 The largest discount we can give for our A series products is 40% due to price restrictions.

A系列產品因為有價格限制，所以我們可給的最大折扣為40%。

例 A few price restrictions are imposed on commodities such as gas, gasoline and electricity.

像是天然氣、石油和電力這些商品就會有一些價格限制的管控。

情境三

原廠所設的規範與限制

例 The violation of any of these restrictions will lead to the cancellation of orders.

若有任何違反這些限制的情事發生，則會取消訂單。

例 Some Products are subject to restrictions on resale.

有些產品有轉銷售的限制管制。

例 We have our restrictions prohibiting distributors from displaying the price for the product on their web pages.

我們有限制經銷商不要將此項產品的價格列在其網頁上。

例 We're considering to add new restrictions to sales of our products in the US.

我們正在考慮要對我們產品在美國區域的銷售加上新的限制條件。

術語直達車，專業補給站

對於任何促銷特惠活動的規劃，除了要設計出吸睛的好康內容與簡潔有力的活動標題之外，同時也要抽絲剝繭地訂出完整的促銷內容，設定出促銷所要針對的人事時地物，以及所要適用的條件。而既然訂出了適用的條件，換言之，也就訂出了不得適用的限制條件了。

一定要有限制條件嗎？就算是十分「不設限」的創意促銷折扣，例如Waitrose超市推出的新興誘人特惠創意：「Pick Your Own Offers」／自己選折扣，讓Waitrose超市的會員自己從近千種的商品中選出十項商品，這十項就可有額外的八折折扣。這樣的不設限，請你猜猜有沒有配套的限制方式？當然有！它的條件可也有個好幾條，例如：

- Open to members only
  只有會員才可享此好康
- Discount applies only to the selected products
  折扣僅適用於特選產品
- The maximum discount available per transaction is £250
  每次交易可扣抵的最大折扣金額為英鎊250元

……最後再加上個限制條件

- Waitrose reserves the right to modify or withdraw this offer at any time
  Waitrose有權在任何時候修改或撤銷此特惠折扣

你瞧瞧，不論促銷活動規模大或小，不論內容、標題多有創意，限制條件也還是有，就算落落長，也是要逐一詳細列出呢！

那原廠要怎麼樣制訂出完整的適用與限制條件呢？此時，請借助基本又好用的5W1H法則，盯住目標與預期成效，逐一來為促銷定調。在這裡，就讓我們跑一回5W1H，一起來將促銷活動的限制條件訂得面面俱到、滴水不漏吧！

What：促銷產品的範圍為何？贈送產品的限制條件為何？

促銷產品的範圍從所有產品、除去某一類型之外的所有產品、某幾類型產品、單一類型產品、新產品、特選產品都有，因此，促銷標的產品一訂定出來，就要將所要排除在外的產品清楚寫明，例如：「Does not apply to A products」／不適用於A類型產品，或是「Limited to B products」／僅限於B類型產品。

對於買★送一的贈送產品，也有會要設限制條件的需要，像是若要求價值須與所購買產品相等或是價值較低，就須這麼說：「Free item must be of equal or lesser value to the lowest priced purchased item.」／免費品項須為與所購買金額最低之產品等值或價值較低的品項。再者，若特惠活動規定的是要贈送免費產品，就會加註「No cash or cash equivalent.」，以限制不得要求改以現金或等價物替換。

Why：特惠活動的目的為何？

一般來說，特惠的目的都是為了要吸引客戶在將要展開的促銷期間多多下單，所以，很明顯的一個限制條件就會是：只有新訂單才適用，既往不咎。對於促銷方案推出前才剛剛下單、訂單還沒出貨的客戶，這點可就讓人很有意見，很想要求原廠通融一下，破個例。因此，這樣的限制條件也是要列出來，像這樣說個明白：「Offer may not be applied to existing, pending or prior orders」／特惠不適用於已下單、尚未出貨或先前的訂單，或是「Offer valid on new orders only.」／特惠僅適用於新訂單。若是特惠活動的目的是要鼓勵網路下單，那就會看到這樣的限制條件：「open to end users placing orders online」／適用於網路下單的最終使用者。

Who：針對的客群為何？

特惠活動的客群資格，一定會列在限制條件中，一定會列在特惠活動網站頁面或文宣資料的中／下方！這個限制條件會決定你看到特惠消息的快樂能不能延續…若經銷商本來看著熱血，但看到「Open to end users only」或「Offer valid for end users only.」，說著僅適用於最終使用者的這個條件時，熱血一定會馬上急凍！大多數會在網站上大書特書的特惠，所針對的客群一定都是最終使用者，而不是經銷商，因為網站的使用者本就是包括一般普羅大眾、一般使用者，所以所暢言的促銷內容多是說給一般客戶所聽。

When：促銷活動舉行的期間為何？

這個訊息在促銷通知的標題中一定看得到，像是促銷活動何時開跑，何時無緣再相見，都得要讓你一看就明瞭，例如特惠期間是至2023年10月31日為止，就可說：「Offer expires on October 31, 2023」，或者要說明特惠的起訖期間，就可這麼寫：「Offer will apply to qualifying orders received between July 1 – September 30, 2023.」，說著在2023年7月1日至9月30日期間所收到的所有合格訂單，皆可適用特惠價。要說特惠，鐵定要將「限時」限得一清二楚，這樣的優惠才有特別，才有吸引客戶快快下單的誘因，不會像是路上偶爾會看到店家所掛「跳樓大拍賣」的布條，一掛就是兩年，看得路人早就疲乏不耐了哩！

Where：促銷活動的地域範圍為何？

特惠方案對地域的限制條件，會有本地市場與海外市場的兩種不同設定，也會有針對不同區域業績狀況而個別設計的特惠，像是「Valid in US & Canada only.」／僅適用美國與加拿大區域，或是「Offer valid for European customers only.」／特惠僅適用於歐洲的客戶。此外，對於「Where」的限制條件，是常會併同「Who」的說明一起出現，例如「Open to end users of the United States only.」，一句就將對象與區域交代完畢！

How：以何種方式來執行？

在「How」之下的限制類別可就多些，都是與客戶要下單時的應用有關。首先，就是特惠折扣多會限制不得與其他折扣併用，像是經銷商已享有經銷商折扣，就不得再將特惠折扣加計應用，限制上就會這樣說：「Discount cannot be combined with any other promotions or distributor discounts.」／折扣不能與其他促銷方案或經銷商折扣併用。此外，原廠可能同時有幾種不同的特惠多頭實施中，因此在限制條件中會要求「You can take advantage of all our special offers, but offers cannot be combined on the same order. 」，不得在同一張訂單中應用一種以上的特惠方案。

再者，原廠可能會要求客戶同一種特惠只能使用一次，此時特惠的限制就會這麼寫：「One time use only.」或是「Customers can use the discount no more than once during the promotion period.」。那下單時要怎麼應用特惠呢？若有折扣碼，就要請客戶務必填上：

「Discount code must be referenced when placing order.」／下單時須加註折扣碼。

當限制條件說得差不多了，到最後還有最後一個限制條件，那就是…保留原廠的最後退路與權利：「We reserve the rights to modify or cancel an offer at any time.」／我們保留在任何時間修改或取消特惠的權利。此外，若要更完整，則會再加上最後的法律防線：「Offer void where prohibited, licensed, or restricted by any laws or regulation or institutional policy.」／凡屬任何法律、法規或機構政策所禁止或以許可規範或限制者，則此特惠方案將屬無效。

Show Time! 換你上場！

5W1H 特惠限制條件整合測驗：請填空，請翻譯！

5W1H	限制條件	例句翻譯
What		僅限於B類型產品。 ________
		免費品項須等值或價值較低。 ________
Why	網路下單	適用於網路下單的最終使用者。 ________
	新訂單	特惠僅適用於新訂單。 ________
Who		僅適用於最終使用者。 ________

When	到期日	特惠至2023年9月30日有效。
		2023年7月1日至9月30日期間之合格訂單可適用此特惠。
Where		特惠僅適用於歐洲客戶。
How	折扣不得併用	折扣不能與其他促銷方案或經銷商折扣併用。
		特惠只能使用一次。
		下單時須加註折扣碼。
其他	原廠所保留之權利	我們保留在任何時間修改或取消特惠的權利。

來對對答案

5W1H 特惠限制條件整合測驗解答：

5W1H	限制條件	例句翻譯
What	促銷產品	Limited to B products.
	贈送產品	Free item must be of equal or lesser value.
Why	網路下單	open to end users placing orders online.
	新訂單	Offer valid on new orders only.
Who	客群	Offer valid for end users only.
When	到期日	Offer expires on September 30, 2023.
	期間	Offer will apply to qualifying orders received between July 1 – September 30, 2023.
Where	區域	Offer valid for European customers only.
How	折扣不得併用	Discount cannot be combined with any other promotions or distributor discounts.
	限用次數	One time use only.
	折扣碼	Discount code must be referenced when placing order.
其他	原廠所保留之權利	We reserve the rights to modify or cancel an offer at any time.

Chapter 2 行銷活動

2-1 參展

Elizabeth: Hello, this is Elizabeth Chou from Blue Tree Healthcare. I'd like to speak with Stephan Kershner, please.

Stephan: Hi, Elizabeth. Good to hear from you. How've you been?

Elizabeth: Not bad! I'm calling to check with you about the Health and Medical Exhibition that we will attend in Seattle during November 19 - 23. Did you receive our notification about the exhibition last Friday?

Stephan: Yes, we did. Sorry for not replying to you!

Elizabeth: It's okay. Will anyone from your company attend this exhibition?

Stephan: I'm interested in going there, but I need to check my schedule and then confirm with you.

Elizabeth: No problem. Please just let me know when you have the answer. If you come, we could arrange a meeting and introduce to you our new marketing tools and strategies to promote our healthcare products.

Stephan: That'll be great! I'll check my schedule and reply to you asap.

Elizabeth: Thanks! I look forward to hearing from you! Bye!

背景說明	
人物	Elizabeth：原廠業務經理 Stephan：經銷商產品經理
主題	原廠參展，邀請經銷商前來參觀： 通知展覽消息 & 順道開個會

譯文

伊莉莎白：哈囉，我是藍樹保健公司的周伊莉莎白，麻煩請找史蒂芬．克什納。

史 蒂 芬：嗨，伊莉莎白，很高興接到妳的電話，妳好嗎？

伊莉莎白：還不錯！我打來是要跟你說一下我們會參加11月19日到23日在西雅圖舉行的健康與醫藥展覽，你有收到我們上個星期五發的通知嗎？

史 蒂 芬：有，我們有收到，抱歉還沒回妳消息！

伊莉莎白：沒關係，你們公司有人會來參觀嗎？

史 蒂 芬：我很有興趣參加，不過我得先查一下我的行程，確定後再告訴妳。

伊莉莎白：沒問題，等你有消息之後再告訴我就可以了。如果你可以來，那我們還可以開個會，跟你介紹一下我們新的行銷工具及策略，以推廣我們公司的保健產品。

史 蒂 芬：太棒了！我會查查我的行程，盡快給妳回覆。

伊莉莎白：好的！期待你的消息了！再見！

通知參展的主旨標題

邀請前來參觀杜賽道夫的PYC展覽，與展望國際公司相見

例句一 Meet Vision International at PYC in Düsseldorf

例句二 Join Vision International at PYC in Düsseldorf

例句三 Vision International will participate at PYC in Düsseldorf

通知參展的主旨標題－加碼通知攤位號

歡迎前來參觀杜賽道夫的PYC展覽，展望國際公司在1202號攤位期待與您相見

例句一 Meet Vision International at Booth #1202 at PYC in Düsseldorf

例句二 Visit Vision International during PYC in Düsseldorf at Booth # 1202!

例句三 Come visit Vision International during PYC in Düsseldorf at Booth #1202!

通知參展消息的內文

若有計劃前來……

例句一 We're looking forward to PYC in Düsseldorf, April 20-23. If you are planning to attend the meeting, we hope you will stop by our booth #1202.

譯文一 期待4月20－23日杜賽道夫PYC展覽的到來！若您計劃前來參加，希望您能到我們的1202號攤位參觀。

可更瞭解產品……

例句二 Vision International will be exhibiting at PYC, April 20-23. Visit us at Booth #1202 and learn about our extensive line of products.

譯文二 展望公司將會參加在4月20－23日舉行的PYC展覽，歡迎您來1202號攤位與我們相見，讓您對我們廣泛多樣的產品有更多的認識。

可另外安排會面時間…

例句三 Vision International is exhibiting at PYC in Düsseldorf. If you are planning to attend, it could be a great opportunity to discuss further our business cooperation. You can visit us at booth #1202, but if you want to pre-schedule a meeting please contact us in advance.

譯文三 展望國際公司將參加在杜賽道夫舉行的PYC展覽，若您有計劃前來參加，這將會是很好的機會，可以讓我們商談進一步的合作。歡迎前來我們的1202號攤位，但若是您想要先安排會面的時間，再請事先與我們聯絡。

提供免費訪客證…

例句四 Vision International will be exhibiting at PYC in Düsseldorf (April 20-23, 2023) and we are offering our customers a FREE exhibition guest pass. PYC 2023 is shaping up to be an exciting event featuring outstanding scientific achievement. We hope to see you there!

譯文四 我們將參加在杜賽道夫舉行的PYC展覽（2023年4月20－23

日），會場上會提供免費的展覽訪客證給客戶。PYC 2023會是一場令人興奮的活動，在展覽中將會為您呈現傑出的科技成就！期待您前來共襄盛舉！

展期全天均可安排會面…

例句五 Vision International will participate at the International Convention & Exhibition Centre, Düsseldorf for PYC 2023. We cordially invite our clients to join us at Booth 1202 for meetings throughout the day!

譯文五 展望國際公司將會參加在杜賽道夫國際會議與展覽中心舉行的PYC 2023展覽，我們竭誠邀請客戶前來我們的1202號攤位參觀，全天均可安排會面。

來攤位拿大獎！

例句六 It’s almost that time again! PYC 2023 is upon us! Come visit us at Booth #1202 to win a fantastic prize!

譯文六 時間又快到了！PYC 2023展覽即將到來！邀請您前來我們的1202號攤位贏得超級好禮！

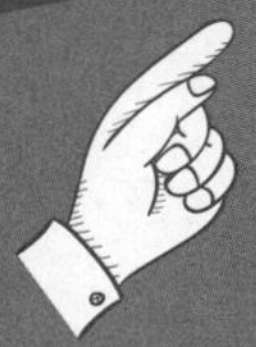

VISION INTERNATIONAL
at PYC 2023

Dear Distributors and customers,

The PYC 2023 is right around the corner! As always Vision International is excited to meet you all in person!

PYC 2023 will be held at the International Convention & Exhibition Centre in Düsseldorf, Germany. Come meet us in A Pavilion, Hall 3 at Booth #1202. It serves as the perfect platform to meet with clients located in Europe, Asia, Africa and the Middle East.

Vision International is eagerly waiting to meet everyone. To schedule your appointment, contact our Sales and Marking Director, Steve Pan, at span@visionintl.com.tw, or Customer Service at customerservice@visionintl.com.tw.

See you in Germany!

Sincerely,
Jeffrey Fang
Vision International

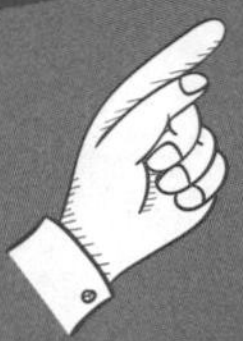

展望國際公司
將在PYC 2023展覽與您相見

各位經銷商與客戶，您們好，

PYC 2023展覽即將到來！展望國際公司一如往常地非常開心能有這個機會與大家相見！

PYC 2023將在德國杜塞道夫的國際會議與展覽中心舉行，請大家前來參觀我們在A館第三大廳的1202號攤位，讓我們有這個絕佳的機會與在歐洲、亞洲、非洲和中東的您們見面。

展望國際公司熱切期盼與您們相見，如要約定會面的時間，請聯絡我們的銷售行銷總監，潘史帝夫：span@visionintl.com.tw，或是與客戶服務部門聯絡： customerservice@visionintl.com.tw。

期待在德國與您相見！

謹上

方傑弗瑞

展望國際公司

Dear distributors,

Vision International will be attending the PYC 2019 in Düsseldorf, Germany, April 20-23, 2023.

We hope you'll visit us at booth #1202 where we will be sharing our expertise on a number of new released research tools and resources.

If you want to pre-schedule a meeting, please let me know what works best with your schedule. The following individuals will be available:

Jason Liu
Marketing Manager
Thursday, April 20 at any time OR
Friday, April 21 before noon

Chris Li
Sales Manager
Saturday, April 22 after 2:30 pm OR

Sunday, April 23 at any time

Please let me know two possible meeting times that are convenient for you so that I can coordinate with their schedules.

Thank you!

Sincerely,

Jeffrey Fang
Vision International

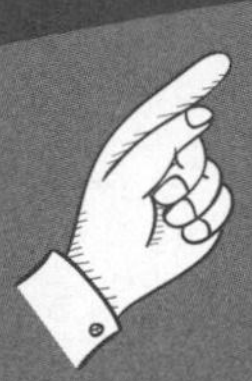

請加入我們
歡迎參加PYC 2023展覽

各位經銷商，您們好，

展望國際公司將會參加2023年4月20-23日在德國杜賽道夫舉行的PYC展覽。

我們希望您能前來我們的1202號攤位，讓我們針對幾項新推出的研究工具，跟您分享我們的專業所知。

如果您想要先安排會面的時間，請讓我們知道哪個時間最能與您的行程配合，我們幾位經理可安排會面的時間如下：

劉傑生

行銷經理

4月20日星期四全天，或4月21日星期五中午前

李克里斯

銷售經理

4月22日星期六2:30pm之後，或4月23日星期日全天

請告訴我您方便的兩個時段，讓我可協調安排這兩位經理的行程，謝謝。

謹上

方傑弗瑞

展望國際公司

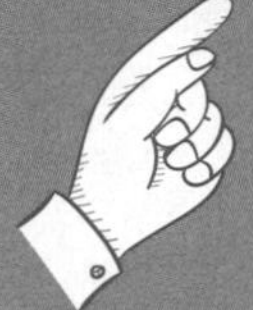

Pioneer Biotech

Meet us in Boston at the Experimental Biology 2023 Boston Conventional and Exhibition Center April 6~9

VISIT US AT BOOTH #1202

Pioneer Biotech will be exhibiting at the Experimental Biology 2023 Meeting at the Boston Convention and Exhibition Center (BCEC), April 6~9. Visit us at Booth #1202 and learn about our extensive line of products!

Come to our booth and get a code for up to 50% discount on ALL enzymes!!

Get ALL our enzymes for up to 50% off prices until April 30 with a coupon code provided at the booth! We look forward to seeing you there!

Can't make it to the meeting? Follow our Facebook and Twitter pages for live updates during the event!

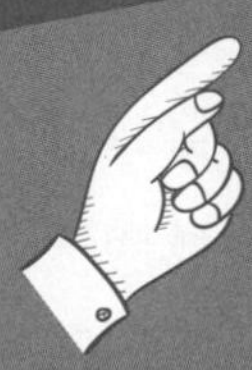

先鋒生技

與您相約在波士頓
實驗生物學 2023
波士頓會議展覽中心
4/6~4/9
歡迎前來 1202 號展覽攤位

先鋒生技將參加在4月6～9日舉行的2023實驗生物學會議，地點在波士頓會議展覽中心(BCEC)。歡迎您來參觀我們的1202號攤位，讓您對我們廣泛多樣的產品有更多的認識！

只要來我們的攤位，就可得到折扣率達50%的優惠代碼，所有的酵素產品都適用！！

使用在我們攤位取得的優惠券代碼，在4月30日之前，購買我們所有的酵素產品，皆能拿到高達50%折扣的價格！期待在會場見到您！

您是否沒辦法前來會場呢？您還是可以從我們臉書與推特上，看到會議期間的實況更新報導喔！

關鍵字急救站—會議

原廠人員能與國外客戶共赴盛會、相見歡的機會，就是參加國際會議、展覽與研討會了。在這些會議與展覽的標題中，除了那各個不同領域與專業的專有名詞與術語之外，剩下來的要角當然就是説明該場集會屬性的關鍵字了，看它是「會議」、「展覽」，還是「研討會」。不過，也請不要以為這些中文的關鍵字有其一對一的英文，對完了就沒事兒了，因為這些詞兒可不是省油的燈，光一個「會議」的中文詞兒，可也有好幾個不同會議種類的英文名稱呢！那分出那麼多種類，到底彼此之間不同在哪裡呢？我們就在這裡把它們好好説一説，一次看個夠吧！

先來看個簡單的總表：

會議種類			
Meeting	會議	Assembly	集會
Conference	會議	Convention	會議，大會
Congress	代表大會	Colloquium	學習報告會
展覽種類			
Exhibition	展覽	Exposition	博覽會
Fair	展覽	Show	秀，展覽

討論會種類			
Seminar	專題討論會	Workshop	專題討論會，研討會
Symposium	討論會，座談會	Panel Discussion	討論會
Forum	論壇		

接下來，我們再來細看各個名詞詞彙的定義囉！

Meeting [ˋmitɪŋ] 會議；集會

an occasion when people gather to discuss things and make decisions, either in person or using phones, the Internet, etc.

人們為了各種不同的目的齊聚而有的聚會，都可稱為meeting，不管是何性質、內容，規模是國際還是區域，時間是定期還是不定期，有規劃與臨時舉行的聚會，還是面對面、電話上或網路上所開的會議，都在meeting這個會議通稱的大範圍裡。另外，我們會看到許多學會或協會每年舉行的年度會議就稱為Annual Meeting，簡稱年會。

Assembly [əˋsɛmblɪ] 集會

a meeting of people who represent different parts of a large organization

Assembly是一個協會、俱樂部、組織、或公司的正式全體集會，參加的人以組織內的成員為主，多有固定的集會時間與地點。

Conference [ˋkɑnfərəns] 會議，討論會，協商會

a large meeting, often lasting a few days, where people who are interested in a particular subject come together to discuss ideas

任何機構組織想要討論、辯論、交換意見、傳達訊息或針對某一課題發表、傳達意見，都可來辦上一場conference。

Convention [kənˋvɛnʃən] 會議，大會

a meeting that a lot of people belonging to a particular profession or organization go to in order to discuss things

任何機關團體組織為了特定的目的與情況進行商討，為了達成共識，而對其成員召開的會議。

Congress [ˋkɑŋgrəs] 代表大會

a formal meeting or representatives, for example from different nations or scientific organizations, to discuss ideas, make plans, or solve problems

此為定期舉行的會議，由各團體派出正式代表與會，人數可有數百人，甚至多達千人，在規模上比conference還大。例如International Congress of Nutrition／國際營養大會。

Colloquium [kəˋlokwɪəm] 討論會；學術報告會

a large meeting to discuss something, usually an academic subject

通常是針對學術研究方面所開的討論會議、座談會，也有人稱其為學術報告會。

Colloquim其實也就是正式的seminar，說到seminar，這類相似的討論會、座談會，可也有好幾個呢！那就讓我們繼續看下去囉！

Seminar [ˋsɛmə,nɑr] 專題討論會

a meeting at which a group of people discuss a subject

Seminar指的是一群人針對一個主題進行討論，提出各自的觀點，通常是指講授課程。

Workshop [ˋwɝkʃɑp] 專題討論會，研討會

an occasion when a group of people meet to learn about a particular subject, especially by taking part in discussions or activities

Workshop指的是較小型的會議，互動性高，通常會有實際展示或有實際動手做的練習機會，也稱為工作坊、研習坊。

Symposium [sɪmˋpozɪəm] 討論會，座談會

a meeting where experts discuss a particular subject

Symposium這個字原指「在一起喝酒」之意，古代希臘人有一面喝酒，一面討論文學、哲學、政治等的習慣，而古希臘哲學家柏拉圖有一篇作品即名為〈Symposium〉，中文譯名包括有〈會飲篇〉、〈饗宴篇〉和〈宴話篇〉，其為一篇對話式的作品，以演講和對話的形式寫成，討論愛的本質。Symposium此字之後用於強調知性談話的內容，如今則用來指稱討論會、座談會。

Panel discussion 專題座談會

a discussion carried on by a selected group of speakers before an audience

Panel這個字在此處為「a group of people who make decisions or judgments」，是小組的意思，可以有專門小組、裁判小組、專題討論小組等，以針對某一個問題來進行公開的討論，例如原廠可通知客戶說「A panel discussion will be held on the topic "Current economic climate"」，說明將會舉行一場針對「經濟景氣現況」主題所做的專題座談會。

Exhibition [ɛksəˋbɪʃən] 展覽，展示會

a public show where art or other interesting things are put so that people can go and look them

The Global Association of the Exhibition Industry（國際展覽業協會，簡稱UFI-稱為Union des Foires Internationales）所定義的「展覽」是一種市場公開活動，在特定期間裡，由許多家廠商齊聚一堂，將各家的產品與服務陳列展示給大眾。Exhibition是最常被應用的展覽屬性名稱，多指具有貿易性質的專業展覽。例如International Exhibition on Food Processing and Packaging Technology／國際食品加工設備暨食品包裝工業展。

Exposition [ˌɛkspəˋzɪʃən] 展覽會；博覽會

a public event or show of industrial products or technology

Exhibition 的法文就是Exposition，應用上指的多是大型博覽會，由數

個展覽（exhibitions）組成，例如2018 Taichung World Flora Exposition／2018 台中世界花卉博覽會臺北國際花卉博覽會。

Fair [fɛr] 展覽

an event where people or companies bring their products for you to look at or buy

Fair一字指的是市集這類傳統型態的展覽會，展覽內容可以包羅萬象，多以展覽及販售為目的，例如Frankfurt Book Fair／法蘭克福書展、International Antiques Fair／國際古玩展、International Hardware Fair Cologne／科隆五金展。

Show [ʃo] 秀，展覽

an occasion when a collection of things are brought together for people to look at

Show是貿易展覽會的意思，像是Copenhagen International Fashion Fair／哥本哈根國際服裝展覽會、Taiwan Hardware Show／台灣五金展、Taipei International Baking Show／台北國際烘焙暨設備展、Tokyo International Gift Show／國際禮贈品展。

術語直達車，專業補給站

展覽所屬的產業名稱叫做「MICE」…… 「MICE」？這不是老鼠「mouse」的複數嗎？是的，當你想到這兒，會這樣自問時，那你也就記下了「會展產業」的頭字語就是「MICE」！我們且來看看這幾隻「老鼠」的身分來歷吧！

M：Meeting／會議
I：Incentive／獎勵旅遊
C：Convention／國際會議
E：Exhibition／展覽

會展產業（MICE）包含了這四項展演活動類型，而「會展」二字指的即為「會議」與「展覽」，其產業範疇內也包含了會議與展覽籌辦過程中所涉及的 「活動」、「旅遊」及其衍生行為。會展產業整合了生產、行銷、餐飲、觀光等產業特性，提供了國際技術、文化、學術交流與合作的平台，是無國界的整合行銷服務。會展產業在歐美各國已有百年以上的發展歷史，近年來，亞洲鄰近國家如日本、韓國、新加坡及中國等，也都將會展產業列為積極發展的重點產業。有「會展亞洲盃」之稱的「亞洲會展產業論壇」（Asian MICE Forum, AMF），也在2015年9月於台北國際會議中心盛大舉行。行政院觀光推動委員會也設有MICE小組，定期邀集經濟部、交通部、外交部、陸委會、經建會、會展公協會及旅遊公會，協調解決會展發展所面臨之問題，希望能在政府的積極推動下，全面提升台灣舉辦會展之國際地位，進而成為亞洲會展的重鎮。

會展產業每年在全世界製造上億元的商機，參展是「面對面行銷」，它有其他行銷工具所無法取代的特性。那企業參展有沒有為自己帶來利多？有沒有達到預期的成效與收益呢？這就有趣了，企業常有為了重要活動而戮力籌劃了幾個月的前置工作，那活動一結束了呢？「終於辦完了，不就該好好放鬆一下嗎？」是啊！不過，要是沒有好好在參展前設立目標，並在參展後評估、追蹤，那參展這件事，能百分之百確定的就只有這件事：你一定會拿到帳單，但不見得能賺到足夠的訂單！

參展一定要預先擬定出目標，有了目標，知道要往哪個方向努力之後，也才能有獲得成效、達成目標的機會。那麼，參展理應要幫企業達成哪些目標呢？首先當然要先來看看大家心中一定念茲在茲、鐵定一想就有的目標：銷售業績！

參展當然不是玩玩而已，首要目標當然一定要獲得銷售實績，那麼，企業花了大把的錢參展，會希望能在銷售方面達到什麼樣的目標呢？在此列出幾個重要的目標，帶你順一下思路、想個透徹囉！

找出有望客戶／Generate sales leads

參展的主要目標之一就是找出新客戶！這些客戶是自動送上門的嗎？是也不是，這些新的有望客戶確實是自個兒走來參觀攤位，表示興趣，但是要吸引他們過來，那就是攤位的設計與行銷策略起了效用之故。參展或刊登廣告的目的就是要藉由「昭告天下」的形式來吸納sales leads，讓有興趣的新客戶能夠循線找上門來，主動與我們連絡，讓企業得以拓展客戶基礎，新增客戶名單。

原廠在參展中所取得的可貴sales leads，在展覽結束後即須有系統地轉給海內外經銷商，請其追蹤，與新客戶聯絡，而在Email通知上就可以這麼說：

例句一 Attached are new sales leads for you. These researchers visited our booth at the recent SLAS (Society for Laboratory Automation and Screening) meeting in Orlando, Florida. Please contact and send them catalog or the specific requested literature.

譯文一 附上新的有望客戶資料如附，這些研究人員參觀了我們最近在佛羅里達奧蘭多SLAS（實驗室自動化與篩選協會）會議上的展覽攤位，請與他們連絡，提供型錄給他們，或是提供他們所要求的特定產品資料。

例句二 Please see the leads from the recent Experimental Biology meeting in San Diego. Please contact them with the information they requested and keep me informed of the success of the leads.

譯文二 請見我們最近從聖地牙哥的實驗生物學學會展覽中取得的有望客戶資料，請與他們連絡，提供他們所需要的資訊，若有成功案例，也請讓我們知道。

直接成交／Make direct sales

企業在展覽中若能施展行銷能力，搭配展覽的特惠或促銷方案，即可吸引客戶在重重好康的誘因之下，當場下單，直接成交！這在參展期間可是大快人心的興奮時刻！而這部分的成效最好評估了，整理一下orders taken & revenue，計算一下所取得的訂單收入，就可知道成效大小，再與原先的估計比較一下，並整理一下客戶的反應，所得出的資料就是可用來設計與修正下次參展行銷策略的絕佳素材囉！

建立聯絡資料庫／Build a contact database

客戶在哪，生意就在哪！要拓展業務規模，就要擴大客戶基礎，而從參展這種撒網方式下手，就是要吸引新客戶，因此，「Number of new contacts」的多寡，增加了多少新客戶，就會是參展成效的一大指標。我們來看一下在會後要檢討說明這一項目標的設定與成效結果時，會是怎麼說囉：

例句　Our goal was to capture at least 100 names of the many visitors we spoke with at our booth and we met that goal with 120 names.

譯文　我們的目標是要從到我們攤位的許多訪客中，拿到至少100個客戶名單，結果我們有達到目標，共拿到了120個客戶名單。

說到新客戶的聯絡資料庫，我們也就順道來說一下所謂的客戶資料庫／customer database。拜現今的全球化與資訊科技之賜，企業的客戶資料在數量上與細目資訊上都比以往多上許多，也更加完整，更能讓我們好好利用，像是做出**「客戶分群」（Customer Segment）**，以做為行銷的依據，或是推行**差異化行銷（Differentiated Marketing）**。此外，也可運用客戶的歷史採購資料，做出**客戶行為分析（Customer Behavior Analysis）**，藉以設計出有效的行銷策略。例如，原廠將新產品資訊提供給經銷商，請經銷商將新產品介紹給適合的客戶時，就可這麼說：「Attached is our most recent newsletter announcing a few new products. Let me know if you have any questions regarding these products or our newsletter. Feel free to promote new products as you see fit for your customer database.」

參展所要追求的除了銷售面的目標之外，最主要的還有**建立品牌（brand building）**這件事。在行銷管理上，品牌的建立與經營可是非常重要的事，關係著公司的獲利能力與未來的成長。藉由參展，企業可以展現出想要形塑的**品牌形象（brand image）**，或是企業已轉型，即相當需要在展覽中昭告天下，**重新定位其品牌（re-position a brand）**，而利用在展覽中的曝光機會，則得以**擴大品牌知名度（brand awareness）**。這些跟「brand」有關的詞語搭配，在談行銷策略或是說明行銷活動的考量因素時，可是個個都經常出現，一定會讓你看得熟到心坎裡呢！另外，brand 的詞語搭配還有一個常客：**「brand loyalty」**，品牌忠誠度，是「the tendency of customers to keep buying the same brand of a particular product instead of trying other brands」，一個廠商若是擁有許許多多具品牌忠誠度的死忠客戶，那可真的會讓廠商感動又感激呢！

在另一方面，透過參展的活動還可讓企業來實地做個市場調查（market test），讓新的行銷策略試試水溫，讓企業研究一下行銷活動的成效，進而依據客戶反應來調整行銷策略。

每年所參加的會議與展覽都是企業的年度大事，既然是大事，就要好好辦、仔細辦，在事前詳細地規劃，在展覽期間有效執行與掌控，再加上事後有系統地追蹤與分析，必定能使參展的成效極大化，讓企業能夠獲利、成長，得以永續經營！

填空

來試試看你是不是對這些會議展覽的名稱都已熟透了呢？

會議種類			
M__________g	會議	A__________y	集會
C__________e	會議	C__________n	會議，大會
C__________s	代表大會	C__________m	學習報告會
展覽種類			
Exh__________n	展覽	Exp__________n	博覽會
F__________	展覽	S__________	秀，展覽
討論會種類			
S__________r	專題討論會	W__________p	專題討論會，研討會
S__________m	討論會，座談會	P__________l D__________n	討論會
F__________m	論壇		

簡答

請問「會展產業」包括了哪四項活動（中英文皆須列出）？英文簡稱為何？

翻譯

❶ 展望公司將會參加在4月20－23日舉行的PYC展覽，敬邀您前來1202號攤位與我們相見，讓您對我們廣泛多樣的產品有更多的認識。

Vision International will be ___________ ___________ PYC, April 20-23. Visit us ___________ ___________ # 1202 and learn about our extensive line of products.

❷ 展望國際公司將參加在杜賽道夫會議與展覽中心舉行的PYC 2023展覽，我們竭誠邀請客戶前來參觀我們的1202號攤位，全天均可安排會面。

Vision International will ___________ ___________ the International Convention & Exhibition Centre, Düsseldorf for PYC 2023. We ___________ ___________ our clients to join us at Booth #1202 for meetings ___________ ___________ ___________!

❸ 我們希望您能前來我們的1202號攤位，讓我們針對幾項新推出的產品，跟您分享我們的專業所知。

____

____

❹ 附上給您的新有望客戶資料如附，這些客戶參觀了我們最近參展的攤位。請與他們連絡，提供他們所需要的資訊，若有成功案例，也請讓我們知道。

____

____

____

____

來對對答案

填空

請往前翻出本單元所列的整理表囉！

簡答

會議 Meeting

獎勵旅遊Incentive

國際會議 Convention

展覽 Exhibition

簡稱「MICE」

翻譯

❶ exhibiting at；at Booth

❷ participate at；cordially invite；throughout the day

❸ We hope you'll visit us at booth #1202 where we will be sharing our expertise on a number of new released products.

❹ Attached are new sales leads for you. These customers visited our booth at the recent exhibition. Please contact them with the information they requested and keep me informed of the success of the leads.

Chapter 2 行銷活動

2-2 支援經銷商參展

Robert: Hello, Jenny. It's Robert from Rich Technology.

Jenny: Hi, Robert. How've you been?

Robert: Just fine! I would like to discuss the conferences and exhibitions you're going to attend this year with you.

Jenny: The exhibition that we've already confirmed to attend is The Joint Annual Conference of Biomedical Science.

Robert: Got it! I know Symposium on Stem Cell Research will be held in June in your country, right? I wonder if you are going to participate in this symposium.

Jenny: I know that symposium... Hmm, we've never attended that symposium before but we could evaluate its cost and effect and let you know our decision.

Robert: OK. You know we aim to develop the market of stem cells this year. We hope you can participate... and maybe we could co-exhibit with you.

Jenny: That's a good idea. I'll go check the related information about this symposium and then discuss the details with you about co-exhibition.

Robert: Good to hear that! So I'll wait for more news from you.

Jenny: Not a problem!

背景說明	
人物	Robert：原廠行銷經理 Jenny：經銷商產品專員
主題	原廠詢問經銷商參展計劃： 經銷商回報所要參加的展覽 & 原廠提議共同參展

譯文

羅伯特：妳好，珍妮，我是理奇科技的羅伯特。

珍　妮：嗨，羅伯特，你好嗎？

羅伯特：還不錯！我想要跟妳討論一下你們今年要參加的會議與展覽。

珍　妮：我們已經確定會參加的有生醫科學聯合年會。

羅伯特：瞭解！我知道在你們國家六月會有一場幹細胞研究座談會，是嗎？你們會參加這個座談會嗎？

珍　妮：是有這一場座談會…嗯，我們是沒參加過這個座談會，不過我們可以去評估一下它的成本與效益，再告訴你我們要不要參加。

羅伯特：好的，妳知道我們今年的主力就是要發展幹細胞這塊市場，我們希望你們能參加…可能我們也可以跟你們一起來辦。

珍　妮：這主意不錯！我會去查查這場座談會的相關資訊，然後再跟你討論一下共同參展的細節。

羅伯特：很高興聽到妳打算這麼做！那我就等妳的消息囉！

珍　妮：沒問題！

說三道四，換句話試試

例一｜詢問經銷商的參展計劃 & 有什麼展覽呢？

例句一 Would you please let me know the exhibitions you will be attending this year?

譯文一 能否請您告訴我您今年會參加哪些展覽？

例句二 I would like to know about the conferences and exhibitions you plan to participate. If there are conferences you have doubts about, please share the related information with us and we will be happy to give you suggestions.

譯文二 我想知道您計劃參加的會議與展覽有哪些，若是有的會議是您不太確定要不要參加的，就請告訴我們相關的資訊，我們很樂意為您提供建議。

例句三 Please inform us about all the exhibitions at your region in the field of automation and control. If there are any exhibitions or conferences you think might be relevant, please also let us know.

譯文三 請告訴我們在您區域內有關自動控制的所有展覽，若您認為有哪些展覽或會議可能有相關，也請讓我們知道。

例二 I 建議經銷商參展

例句一 I know you decided not to participate at PAC conference. I checked this conference once again and it looks very interesting and relevant to our business. Could you consider again your participation at this conference?

譯文一 我知道您決定不參加PAC會議，我有再查了這個會議，發現它讓人滿感興趣的，也與我們的業務有關，您可不可以再考慮一下參加這個會議呢？

例句二 I know you didn' t include this convention in your plan this year, but I wonder if you would be interested to exhibit - if we will share the costs with you.

譯文二 我知道您沒有此會議的參展計劃，不過，如果我們跟您分攤費用，不知您會不會有興趣參加。

例句三 Can you give me some information about PAC conference? Are you going to participate in this conference this year?

譯文三 您能不能給我PAC會議的一些資料？您今年會要參加這個會議嗎？

例三 | 與經銷商確認欲參展的展覽

例句 Please check the attached table and confirm whether these are the exhibitions you will exhibit. If there is any information missing, please add it to the list and then Email to me.

譯文 請看附件的表格，並請確認這些是否就是您今年會要參加的展覽，若是漏了任何的資料，就請您加到清單裡，再Email給我。

例四 | 詢問經銷商需要哪些文宣資料

例句 I know that you will be exhibiting at the PAC Conference. What marketing materials will you need? Please share with me whether you have any particular needs for this conference.

譯文 我知道您將會參加PAC會議，您有需要什麼樣的行銷資料嗎？若您對此次參展有任何特別的需要，就請告訴我。

關鍵字急救站

我們在這兩個單元說了那麼多參展的事，你記不記得我們說「參」展的這個「參加」，前前後後用了哪幾個不同的動詞呢？那個…有attend、participate，還有join！非常之好！那這些字都是「參加」的意思，所以可說通通都一樣，絕對都可以通用囉？如果有人拿這些「絕對」、「一定」的論述來考你，那麼那個論述絕對是有問題，一定是錯的啦！這些動詞確實都可用在「參加」展覽、會議、討論會上，但在定義上還是有些不同，現在就讓我們靜下心來看看這些「參加」的定義、用法與搭配詞，另外再大放送，加上同樣是「參加」一族的take part in和enter囉！

attend → 靜態、被動參與，正式

英英解釋：to be present at an event or activity

Attend為出席、參加某件事情或活動的意思。常見的搭配詞有attend a meeting（會議）／training（訓練），attend 為正式的表達方式，常用於形容參加正式活動，或經常去。不過出席是出席了，並不代表會積極投入活動中。

在分析展覽的登記人數時，就會用到attendee（出席者）這個字。

例句　A great location, excellent hosting, and perfect weather lead to the largest registration turnouts: 1925 registered attendees.

譯文　因為地點很好、展覽辦得很好、天氣又超好，所以登記的出席人數是史上最多的一次：有1925位。

另外，片語attend to的意思跟attend一點也不一樣！attend to有「處理（事或人）」的意思，像是說著「We will attend to your request as quickly as possible.」／我們會盡快處理您提出來的要求；此片語也有「為（顧客）服務」的意思，而在正式的用法上，則有「注意，傾聽」之意。

participate → 主動、積極參與，較正式

英英解釋：to take part in something

Participate 是指參與某活動，並不包括加入組織之意，常與介詞in連用：participate in 例如，participate in the survey (調查)，為比較正式、書面的表達方式。

take part in → 主動、積極參與

英英解釋：to be involved in an activity with other people

Take part in 一般指的是參加群眾性的活動，在其中身體力行，盡一己之力，例如 take part in a revolution （革命）/ contest（競賽）/ struggle（鬥爭）等，為較常用的口語表達方式，亦可在稍微正式的場合使用。

join → 參加已存在的活動或團體，成為其中一員

英英解釋：to become a member of an organization, club etc.

Join 為及物動詞，指參加一群人的活動，或是加入某個組織，並成為其中一員。常見搭配詞有：join team（隊）/ union（協會）。

Join in 指的是參加某項活動或運動，to do an activity with people who are already doing it。

Take part in 和 join in 常可換著用，例如take part in / join in the

discussions 一起參加討論，不過，若是指參加組織，成為其中一員，則會用join in。另外，當我們說join in something 時，該活動都是已經開始、正在進行中，這個時間點也是區分take part in 與join in 的一個因素。

enter → 參加已存在的活動或團體，成為其中一員

英英解釋：to become a member of or start working in

常見的搭配詞有：enter the army（軍隊）/ political party（政黨）。

術語直達車，專業補給站

在參展這個行銷大代誌上，原廠有兩個方向可施力，一個是原廠自己參展去，另一個則是由經銷商參加該區域的展覽，此時是經銷商在替原廠打天下，所以原廠應要提供支援，給予足夠且有力的彈藥，讓經銷商好好地替原廠打響名號、磨亮品牌、賺進業績，締造雙贏佳績！

那麼原廠支援經銷商參展有哪幾種方式呢？從折扣援助、物資支援，到人員親上火線都有！在這兒就讓我們一一來瞧瞧要施展行銷大法時，有哪些點可施力，有哪些方向可以幫助經銷商擴展業務，讓原廠贏得面子又贏得裡子囉！

折扣援助

經銷商參展時提供給客戶的協助，最乾脆也最容易吸納業績的來「攤」好禮，一般說來就是折扣優惠了！

折扣券／Coupon

在經銷商參展時，經常會要求原廠提供促銷折扣券，以吸引客戶前往參觀展覽，並吸引客戶在展覽中或展覽結束後的一段期間內下單。

例句 We will provide you coupons which offer 10% discount valid for six weeks after exhibition.

譯文 我們會提供10%折扣的優惠券給您們，可在展覽結束後六個星期內使用。

特別折扣／Special discount

有些展覽的特惠是促銷某類產品，吸引客戶在展覽開始後的一段時間內下單，以享特別折扣的優惠。下面的這一個範例，就是原廠願意提供折扣而給經銷商的一個正向且百分百支持的回覆：

範例 We could offer you a 15% special discount for our kit products from July 1 to December 31, 2023. We believe this upcoming exhibition is a good opportunity to promote our kit products to your customers and you could rest assured that you'll get our full support!

譯文 針對我們的套組產品，我們可提供15%的特別折扣，適用期間從2023年7月1日到12月31日為止。我們相信即將來到的這場展覽是我們促銷套組產品的一個好機會，您可以放心，我們將會全力給您支持！

物資支援

原廠會給予經銷商的物資支援指的就是行銷用的各種素材與物品，也就是 promotional items 或 marketing materials，這類物資主要有兩大類，一類是文宣，另一類就是贈品了。我們先來看看文宣有哪幾種不同的類別：

中文名	英文名	中文名	英文名
型錄	catalog	小冊子	brochure
印刷文宣	literature	傳單	flyer
資料夾	folder	海報	poster

接著，我們再來看看贈品的型式。贈品可不像文宣型式那麼客氣了，贈品雖不是多到像牛毛那麼多，但也確實沒個範圍，要有多少種就有多少種哩！在此我們將常見的幾種贈品列出來讓參觀者看得高興囉！

中文名	英文名	中文名	英文名
原子筆	ball-point pen	磁鐵	magnet
便利貼	post-it	水壺	water bottle
輕便袋	tote bag	填充玩偶	stuffed doll
T恤	T-shirt	棒球帽	baseball hat

經銷商有要參加會議或展覽，那原廠可提供哪些文宣給經銷商呢？這是需要詳實溝通一下的。從下列的範例，就可看看原廠可以如何要求經銷商提供資訊，如何配合囉！

範例一 Please send me the link to the conference so that we can review its program to make sure what the appropriate literature and catalogs are.

譯文一 請給我此會議的網頁連結，這樣我們就可看一下議程，好確定什麼樣的印刷文宣與型錄會適合此會議。

範例二 For the upcoming meeting, we'd be happy to supply you with any posters and literature you want, and can send the items with your order shipment, or separately if needed. A full listing of our current posters & flyers is available at our website. If you'd like to order any of these items, please feel free to Email me directly and we'll ship them to you quickly.

譯文二 對於即將到來的這場會議，若您有想要任何的海報與印刷文宣，我們都很樂意提供給您，可跟著您的訂單一起寄給您，若有需要的話，也可單獨寄出。我們目前可提供的海報與傳單的清單，已放在我們的網站上了，若是您有要訂購任何品項，請隨時發Email給我，我們將盡快為您寄出。

在提供展場贈品這部分，原廠一般都會提供好些免費的小東西，送給經銷商，讓他們在展場上利用，衝些人氣，提升來客開心指數。若是經銷商要求更多的量，那可能就不好意思啦！恐怕要收費囉！請來看看下面這一則範例，說的是經銷商請求支援，原廠爽快配合，同時原廠也說明了好意支援有個限度，行銷支出也有個控管的機制，所以，若是經銷商還要更多的話，會是如何來計價。

範例 We would be happy to work with you to provide updated flyers and posters. Also we could provide T-shirt (Qty. 50) and magnets (Qty. 200) for free for you to distribute at the upcoming exhibition. If you want more than what we are offering, then each T-shirt will cost US$10 and magnet US$ 1 each (not including shipping fee).

譯文 我們很樂意跟您合作，我們會提供新版的單張型錄與海報，另外我們還會提供免費的T恤（50件）和磁鐵（200個），讓您在即將舉行的展覽上分送。若是您還需要更多的量，則每件T恤的成本為10美元，每個磁鐵的成本為1美元（不含運費）。

此外，原廠其實也會想知道經銷商會是怎麼利用所提供的贈品，同時，在經銷商告知贈品活動的方式之後，其實原廠更可以提供些建議，若是有多搭配些贈品的需要，還可再提供給經銷商，讓活動可辦得更加「澎湃」。那原廠要怎麼樣問起呢？請看看囉！

範例 Please let me know your marketing plan as to how you will use these promotional items at the exhibition. Would attendees need to take any specific action (i.e. take a survey, speak to a Sales Representative, etc.) in order to receive any of these promotional items? If you have a specific promotional plan to use these items to generate booth traffic or to reach a certain goal, please let me know. I may have other items I can suggest that would work well for your purpose. If you have items of other suppliers as well, please let me know what they are so that I won't suggest an item you will already have on hand.

譯文 請讓我知道一下您們的行銷計畫，告訴我您們在展覽上會是怎麼運用這些促銷贈品的，參觀的人有要做什麼樣的動作（例如填調查表、跟業務代表洽談等等）才可拿到這些促銷贈品呢？如果您們有設計了特定的促銷計畫來用這些贈品，以增加攤位參觀流量，或是達到某個目標，就請讓我知道一下，我可能還可針對您們的目的，給個搭配品項的建議。若是您們還有其他供應商的促銷贈品，也請告訴我有些什麼，這樣我就不會再去提供那些您們手上會有的贈品了。

原廠若能對經銷商所要進行的活動多一些瞭解，就能提出更切合活動目標之專業且有效的建議。原廠支援夠力，經銷商參展成功，衝出漂亮的成效與業績表現，那就是原廠與經銷商共同努力的結果，雙方都能共榮共贏且共樂呢！

原廠人員支援前線

經銷商參展，原廠最大的誠意就是有原廠人員親自動身出馬！這出馬的人員可是各號人物都可能會有，包括業務、技術、行銷推廣等部門的主管與人員，以藉著參展機會，好好與經銷商來個相見歡，一起辦好參展大事，一起刺激業績表現。

那原廠人員的支援型態有哪幾種呢？來攤位站台，讓經銷商、也讓來訪客戶感受到原廠跨海支援的重視程度與盛情，這是一定會有的會場現場支援。原廠的業務及行銷人員在現場可提供更完整的產品或研發等訊息，而技術人員可協助進場設備架設作業，也可立即為經銷商或客戶解答產品問題，這些都是原廠人力大有功效的地方。而在會場與客戶單獨面對面討論，也可以有幾種不同的方式，例如：

- 租會議室，單獨會面
  Rent a room for private customer meetings
- 安排在攤位上與客戶會面
  Set appoints with customers at booth

除此之外，原廠支援人力還有可能可以順道提供下列的大利多，增加客戶來參觀展覽的意願，同時也可提升品牌的知名度與形象！

辦講座／Have a Seminar

原廠人員都已經飄洋過海來支援，當然要盡可能地發揮其功效，所以，光是在攤位站台相挺還不夠，通常會來辦個教育訓練，有的還會在展覽會場租用會議廳，辦個講座，以能將原廠產品的發展與應用等資訊，有效且深入地當面介紹給當地的客戶。我們來看一下要辦講座會討論到什麼樣的訊息：

範例一 We suggest to organize a seminar in the conference. Dr. Chen, our Technical Manager could deliver a presentation about our new products.

譯文一 我們建議可在會議中辦一場講座，我們的技術經理陳博士可以就新產品來做個簡報。

範例二 About the discussed keynote speech, it'll be me giving the talk and I will get back to you with an abstract. Please let me know if there is any word limitation on the abstract.

譯文二 關於所討論的專題演講，會是由我來主講，之後我會提供摘要給您，至於摘要有沒有什麼字數限制，就再請告訴我一聲了。

安排客戶拜訪／Make customer visits

原廠派人員支援，除了辦講座之外，通常都會有這個必要的行程：客戶拜訪。客戶拜訪所安排的客戶都是對經銷商、對市場來說屬於重要或具指標性的客戶，而原廠人員親自登門拜訪，客戶也會覺得備受重視，最重要的是能夠讓他們面對面地就產品與需求等事項來與原廠討論與協商。關於客戶忠誠度這件事，除了看產品的品質之外，就端視原廠與經銷商所提供的服務規格與盡心的程度了，而原廠人員親訪，就是提高客戶忠誠度的大好機會！

Show Time! 換你上場！

填空

中文名	英文名	中文名	英文名
型錄		小冊子	
印刷文宣		傳單	
資料夾		海報	

簡答

請寫出本單元説到「參加」的五個英文單字／片語。

翻譯

❶ 能否請您告訴我您今年會參加哪些展覽？

❷ 我們會提供10%折扣的優惠券給您們，展覽結束後六個星期內有效。

❸ 參觀的人有要做什麼樣的動作才能拿到這些促銷贈品呢？

❹ 如果您們有特定的促銷計畫來用這些贈品，以增加攤位參觀流量，或是達到某個目標，就請讓我知道一下，

❺ 我們建議可在會議中辦一場講座，我們的技術經理陳博士可以就新產品來做個簡報。

❻ 關於所討論的專題演講，會是由我來主講。

填空

請往前翻出本單元所列的整理表囉！

簡答

attend, participate, take part in, join, enter

翻譯

❶ Would you please let me know the exhibitions you will be attending this year?

❷ We will provide you coupons which offer 10% discount valid for 6 weeks after exhibition.

❸ Would attendees need to take any specific action in order to receive any of these promotional items?

❹ If you have a specific promotional plan to use these promotional items to generate booth traffic or to reach a certain goal, please let me know.

❺ We suggest organizing a seminar in the conference. Dr. Chen, our Technical Manager, could deliver a presentation about our new products.

❻ About the discussed keynote speech, it'll be me giving the talk.

Chapter 3 業績管控

3-1 業績檢討

Ann: Hello, Ann speaking. How may I help you?

Oliver: Hi, Ann. This is Oliver from Kingdom Enterprise. I'm calling to discuss your sales in the last quarter with you. From the quarter calculation we have noticed that there is a significant decrease at the antibody sales in your region. Could you share with me the reason that might cause this decrease?

Ann: Okay. We just had our internal sales meeting yesterday to review the market status and adjust our marketing strategies. As we reported to you previously, our competitor, TOP Tech, sold their antibodies at extremely low prices, which did affect our market and several key customers chose to buy from them.

Oliver: We did aware of this. We need to develop more effective strategies to raise the sales again. Whenever you find yourself in competition with TOP Tech, please let me know and we'll find a better pricing for each case.

Ann: This will be a good way to better control the cases!

Oliver: We don't want to lose any deal! If there's any tough competition that you need our support, just tell me.

Ann: I'll do so! Thanks for your support!

背景說明	
人物	Ann：經銷商業務經理 Oliver：原廠業務經理
主題	業績表現檢討： 原廠詢問業績下降原因 & 經銷商反映競爭者削價競爭的狀況

譯文

安　：您好，我是安，有什麼需要我服務的嗎？

奧利佛：您好，我是精頓企業的奧利佛，我打來是要跟妳討論一下上一季的銷售額。在算每季銷售額時，我們發現妳們區域的抗體銷售減少很多。妳可以跟我們說明一下是什麼原因嗎？

安　：好的，我們昨天也剛開了我們內部的銷售會議，檢視市場的狀況，也調整了我們的行銷策略。如我們先前跟你報告過的，我們的競爭者頂尖科技以極低的價格在賣他們的抗體，那確實有影響到我們的市場，而幾個重要客戶也轉向他們購買。

奧利佛：我們是知道這個狀況的，我們需要想出更有效的策略，好能把業績再拉起來。當你們再碰到頂尖科技的競爭時，請告訴我們，我們會就各個案子的狀況，給個更好的價格。

安　：這會是個好方法，讓我們對案子可以掌控得更好！

奧利佛：我們不想丟了任何一個案子！若是妳有遇到任何難對付的競爭狀況，有需要我們協助的話，就請告訴我。

安　：我會的！謝謝你的支持！

說三道四，換句話試試

例一｜業績比先前同期差，為啥呢？

例句一 The newly released sales figures show that your sales in the past month dropped by 15% when compared to the results from September 2021. Why do you think sales dropped so dramatically?

譯文一 從剛公布的業績數字看來，你們上個月的業績比起2021年9月掉了15%，你們認為業績怎麼會掉那麼多呢？

例句二 In the first quarter of our 2022 financial year, your sales dropped by 12%, compared with a year ago. We would like to know if the decline is due to competition, ending of major projects or due to the budget cuts.

譯文二 在我們2022會計年度裡的第一季，你們的業績比起去年同期下降了12%，我們想知道下降的原因是因為競爭，或是大案子結束了，還是因為預算刪減呢？

例二｜業績比同期增加耶！

例句一 Your sales volume was 52,800 USD in March, a 4% rise compared to February, and up 20% in value compared to March last year!

譯文一 你們三月的銷售額是52,800美元，比起二月增加了4%，比起去年三月增加了20%！

例句二 Your sales volumes increased by 10% in July compared with the same month a year earlier while we had hoped for a rise of just 5%!

譯文二 你們七月的銷售額比起去年同月增加了10%，而我們原先期望的增加幅度只有5%呢！

例句三 I noticed that, comparing to previous months, the decrease this month was much smaller!

譯文三 我注意到跟前幾個月比起來，這個月落後的程度小多了！

例三 | 業績沒有什麼增長！

例句 I noticed that your sales were flat-lining and competitors were gaining ground on us.

譯文 我注意到你們的業績平平的，沒有起色，競爭者都快追上我們了。

例四 | 同期業績徹底比一比，有高也有低！

例句一 Your sales figures showed a 5.3% increase compared with the first quarter of 2021, but a 1.5% decrease from the same quarter a year ago.

譯文一 你們的業績比2021年第一季增加了5.3%，但是比去年同季減少了1.5%。

例句二 The total amount of your orders in the first quarter of 2023 decreased by 15% in comparison to quarter four of 2022, but was up 27% compared to quarter one 2018!

譯文二 你們2023年第一季的訂單總額，比起2022年第四季是減少了15%，但比起2022年第一季可是多了27%！

關鍵字急救站

要談業績，要檢討銷售狀況，可也要知道業績有哪幾種說法呢！「業績不是就sales嗎？」沒錯，但是sales搭配了其他字詞後，說的也是業績啊！……呵呵，這樣說來還挺無趣的，但我們怎能不識其他的業績說法呢？總是要能夠在說膩、寫膩了sales之後，換個口味吧！那麼，就請開始品嚐享用這幾種不同的味道囉！

Sales

英英解釋：the total number of things that a company sells within a particular period of time, or the money that it earns by selling things

看了這個說明，我們會看到兩個要點，一是sales可指稱數量，也可指稱金額，所以sales可解為銷售量，也可當作銷售額。另一個要點即為要說sales，談業績，一定有時間這個因素，一定會與「期間」緊扣著來說明。

Turnover

英英解釋：the value of the goods and services that a company sells in a particular period of time

Turnover是企業在一段期間內銷售貨物與提供服務所得到的營收價值……也就是營業額。

從sales和turnover的定義看來，兩者指稱的標的都相同，因此也經常交換使用。另外，我們也常會看到這兩個字合用：sales turnover，

「Sales turnover is the company's total revenue, both the invoice, cash payments and other revenues.」，是企業銷售與營運所獲取的收入。那麼，照這樣說來，sales和turnover合體之後是有指什麼特別的東西呢？有的，它們合體之後最特別的地方就是：這詞兒一樣還是指「業績」的啦！

Sales volume

Volume的音標是 [ˋvɑljəm]，請注意一下它的發音喔！Volume是「an amount of something」，是指生產或交易的數額與數量。那sales volume就清楚了，說的就是銷售額、銷售量囉！

Sales performance

Performance是成績、成果、績效、表現的意思，在人力資源管理上，常會看到這件大事：performance appraisal／績效評估，這可是關乎你年終獎金和紅利拿得開心或黯淡的心情指標呢！那performance前頭加個sales，很合理，說的就是銷售的成績與表現，要探討的其實也就是銷售額或銷售量囉！

另外，說到銷售的成果，「成果」這詞兒有沒有讓你想到什麼英文字呢？「當然有囉！……嗯……那個……應該是……我覺得……」，請不要太費神，想一下就快出來了囉！我們來想想「成果」這詞兒有兩個字，可否拆成「成就」與「結果」呢？嗯！我提示的夠明顯了，所以答案就是…achievement和result，所以囉，sales achievement或sales results說的也一樣都是銷售額或銷售量的這些銷售成果喔！

術語直達車，專業補給站

要說業績，要評估銷售表現，免不了一定要跟往事乾杯，拿往事來比一比，看看是只能慨歎地念著當年勇，還是可以驕傲地說現在最勇！要比較，那比出來的結果可是有上有下、有增有減、有高有低、有升有跌、有漲有跌、有起有落…你有沒有讀出我列出這些四字詞的言外之意呢？「啊不就是說業績比較上的增加或減少？」喔不！哪那麼淺薄啊？！我在這字裡行間給的暗示是：中文對「增減」的說法有很多種，英文也是的啦！哈哈！我要說的也沒太深沉，但可是有其含意的呢！好了，那到底英文說法有哪幾種，我們在談業績增減時能夠如何地變化多端，就請各位看官看下去囉！

增加或減少

例句 Your sales volumes __________ by 10% in July compared with the same month a year earlier.（下面所列動詞皆為過去式，因為要檢討的業績都是已經發生的、已經過去的事了。）

- 增加、上升、成長：rose / increased / climbed / grew / mounted up / went up / hiked / shoot up
- 激增、急升：soared / skyrocketed / surged / boomed
- 減少、下降、下滑：decreased / reduced / fell / dropped / declined / went down / dipped
- 驟減、暴跌：slumped / plummeted / plunged / nosedived / tumbled

「plummet」當名詞時的意思為墜子，當動詞的意思為墜落、筆直落下。

「plunge」當動詞時有跳入、急降的意思。

「nosedive」這個結合nose＋dive的複合字指的是飛機俯衝，而dive這個字是潛水、俯衝的意思。

「tumble」當動詞時有跌倒、滾下、墜落的意思，所以這種方式的下降就會是暴跌！

持平、維持不變

例句 Your sales volumes __________ in July compared with the same month a year earlier.

答 remained steady / remained constant / remained stable / stayed constant / were flat-lining / leveled out / leveled off / stabilized

波動

例句 Your sales volumes __________ in the past months.

答 fluctuated / rose and fell

Show Time! 換你上場！

填空

才剛看完前兩頁的一堆增加、減少、持平的動詞，你都記得了嗎？沒關係，我們馬上抓一些來考一考，這樣可以馬上激活一下你的記憶力，讓不太清楚的變清楚，「增加」你說明業績的能力，「減少」動詞用得太過單調的機率囉！

（動詞部份請填過去式）

增加		
r________e	in________	cl________
g________w	w______t u_____	m______u______
減少		
decr________	decl________	f______l
dr________	re________	w______d_____
驟減		
sl________	plum________	plun________
持平		
remained sta______	remained ste______	remained c______
波動		
f________	r______ and f_____	

簡答

請寫出至少四種的業績說法囉！

翻譯

❶ 我們注意到你們區域的業績減少了許多，你能告訴我是什麼原因造成的嗎？

We have __________ that there is a __________ __________ at the sales in your region. Could you __________ __________ me the reason that might __________ this decrease?

❷ 你們上個月的業績比起2021年9月掉了15%，你們認為業績怎麼會掉那麼多呢？

Your sales in the past month __________ __________ 15% when __________ __________ the results from September 2018. Why do you think sales dropped so __________?

❸ 你們的業績比2022年第一季增加了5.3%，但是比去年同季減少了1.5%。

____

____

來對對答案

填空

增加		
rose	increased	climbed
grew	went up	mounted up
減少		
decreased	declined	fell
dropped	reduced	went down
驟減		
slumped	plummeted	plunged
持平		
remained stable	remained steady	remained constant
波動		
fluctuated	rose and fell	

簡答

Sales、turnover、sales volume、sales performance、sales revenue、sales achievement、sales results都是囉！

翻譯

❶ noticed；significant decrease；share with, cause

❷ dropped by；compared to；dramatically

❸ Your sales figures showed a 5.3% increase compared with the first quarter of 2022, but a 1.5% decrease from the same quarter a year ago.

業績管控

3-2 年度檢討報告

Janice: Hi. This is Janice Chen. Is Hank there, please?

Hank: Hank speaking. How are you, Janice?

Janice: Great! I'm calling to make you feel great, too!

Hank: Really? Come on! Tell me. It's a rainy Monday. I do need some good news to cheer me up!

Janice: You're gonna love this news! As the 3rd quarter came to an end, Beacon's sales result for these 3 quarters shows a significant growth of 35%! Well done!

Hank: You know our sales and technical service teams work really hard to increase our customer base! Our strategies did work well!

Janice: We'd greatly appreciate your efforts! In order to motivate your teams, we decided to offer you a bonus! If your sales results exceed your sales target of 2023, we'll give you 50% bonus of the excess of exact sales volume over target.

Hank: Great! That's a generous offer!

Janice: You know you'll always have our full support!

Hank: We'll be making every effort to achieve and exceed the sales target!

<table>
<tr><th colspan="2">背景說明</th></tr>
<tr><td>人物</td><td>Janice：原廠業務經理
Hank：經銷商產品經理</td></tr>
<tr><td>主題</td><td>總業績檢討：
前三季業績表現好 & 原廠再下利多，承諾超標紅利</td></tr>
</table>

譯文

珍妮絲：嗨，我是陳珍妮絲，請問漢克在嗎？

漢　克：我就是，妳好嗎，珍妮絲？

珍妮絲：我很好哇！我打來就是要讓你也覺得很好呢！

漢　克：真的嗎？快點告訴我！今天是下雨天，又是星期一，我真的需要一些好消息來提振精神呢！

珍妮絲：你會愛聽這個消息的！第三季快要結束了，必肯公司這三季的銷售額大大地成長了35%！做得好哇！

漢　克：妳知道我們的業務團隊和技術服務團隊為了要擴大我們的客戶基礎，真的都非常努力！我們的策略成功了！

珍妮絲：很感謝你們的努力！為了要激勵你們的團隊，我們決定提供紅利！如果你們的業績超過了2023年的銷售目標，那我們就會給你們超標部分業績的50%，當作紅利！

漢　克：太棒了！這紅利條件很不錯！

珍妮絲：你知道我們一直都會全力支持你們的！

漢　克：我們會盡所有努力來達成業績目標，並朝超標前進！

一次說清年度檢討與規範　範例

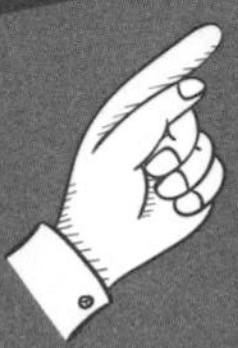

Manufacturer: HongDa Health Enterprise Distributor: Beacon Corp.

Dear Hank,

As 2022 has come to an end, we wanted to write to share with you HongDa's encouraging sales results for 2022. 2022 was another strong year for our company as our overall product sales grew by a robust rate of 36%!

Here is a table showing 2022 worldwide, distributor, your geographic region as well as your own sales growth:

	Q4 2022 Growth	2022 Overall Growth
Worldwide Product Sales Growth	38%	32%
Total Distributor Growth	45%	43%
European Region Growth	35%	37%
Beacon	40%	38%

Distributor sales are an important factor in our company's growth and we would like to continue to try our best to help each distributor capture and benefit from the full potential of our high quality products.

We have many activities planned for 2023 to help continue our strong sales growth. We'll maintain a rapid new product release schedule, have larger overall product offering, and continue our "$150" promotion to encourage new customers to try HongDa's products. We will have quarterly promotions as well. We are also willing to work with you on a country specific promotion tailored to your needs and plans.

We feel that 2023 offers an even better opportunity for us to grow sales as the impact of our large new product increase started in 2022 will really start to take effect in 2023. We look forward to working with your team in 2023 and beyond to help achieve our collective sales and growth targets. We will be in touch with you soon regarding sales targets for 2023 as well.

With Best Regards,

Janice Chen
Sales Manager
HongDa Health Enterprise

一次說清年度檢討與規範　範例　譯文

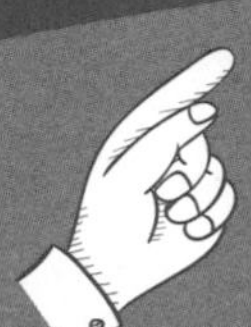

原廠：宏達健康企業
經銷商：必肯公司

漢克，您好，

2022年已將近尾聲，我們要來跟您分享宏達令人振奮的2022年銷售成績。2022年是我們公司再一次的豐收年，我們整體的產品銷售強勢成長了36%！

下表列出了2022年全球、經銷商、您所屬地理區域，以及您公司的銷售額成長資料。

	2022年第四季成長率	2022年全年成長率
全球產品銷售額成長	38%	32%
全球經銷商總成長	45%	43%
歐洲總成長	35%	37%
必肯公司總成長	40%	38%

經銷商的銷售業績是我們公司成長的一個重要因素，我們將會持續盡我們所能，協助每一位經銷商，讓經銷商能夠開發出我們高品質產品的全部潛能，並獲取收益。

我們為2023年規劃了許多的活動，以期持續我們業績的強勁成長，我們在新產品推出計畫上會維持快速的步調，會提供更多的產品，並延續「150美元」的促銷計畫，吸引新客戶試用宏達的產品，我們也會推出每季的促銷活動，另外，我們也願意跟您配合，在您的國家推出符合您需求與計畫的特惠活動。

我們認為2023年對我們的銷量有更好的成長機會，因為我們在2022年增加了許多的新產品，這效應將會在2023年開始顯現出來。我們期待在2023年及未來與您合作，齊力達成我們的總體銷售額與成長目標。我們很快就會再跟您連絡，討論2023年的業績目標。

祝好

陳珍妮絲
業務經理
宏達健康企業

關鍵字急救站—review / forecast

要準備年度檢討與報告，兩大主軸就是掌管回顧、評估、省思的「Review」，以及負責前瞻、預測、預應的「Forecast」了，這兩件事會是年度檢討的關鍵，所以我們就來看看這兩個關鍵字的細部內涵囉！

Review [rɪˋvju]

Review的英英解釋為「the process of studying or examining a situation, policy, or idea again in order to decide whether it is suitable or satisfactory」，意思是要回頭檢視，給過往表現的得失與功過下個定論，是審視，也是評論。

年度檢討報告裡一定會見到「Sales Review」，也就是業務／業績檢討。那該怎麼檢討呢？可以從業績、市場、客戶等方面來談，談的時候要分類、要分析，要拆開來看，也要以總體來說，這樣才能算得上是徹底的檢討。我們現在就來看看可以從哪些點來切入探討年度檢討報告囉！

➡ Forecast Review／預估目標檢討

在每個新年度開始之前，企業一定會訂出業績目標，做出預測，所以在一年結束之後，務必要來檢討一下業績達標與否，看看預計要執行的工作完成了沒。在談業績的這個部分，也不是說說總業績就算了，若要參透數字背後的意思，可得好好分析一番，可以這麼分類來談：

- Sales Review by Month or by Quarter → 以每月或每季分析

- Sales Review by Product Line → 以產品線分析
- Sales Growth Comparison → 業績成長比較

我們在上一個單元有說明了當期業績與前一個月／前一季／去年同期等的比較方式，這時候可就是表格（Table）、折線圖（Line Chart）、長條圖（Bar Chart）、圓餅圖（Pie Chart）等各種分析圖表都要出籠效力的時機了呢！對了，請你看一下Pie Chart的第一個字……沒錯，這個Pie就是我們愛吃的那個「派」！

➡ Critical Opportunity Review／關鍵機會檢討

好好整理一下這個年度裡所發生的重要事件與關鍵機會，檢討一下決策與因應處理的執行效率、適當性與有效性，從其中獲取經驗，找出值得嘉許與可以改進的點，找出successes與things to improve，讓之後有狀況發生時，可以用習得的經驗來處理，以能在關鍵點建立企業進一步發展的機會。

➡ Key Account Review／關鍵客戶檢討

關鍵客戶對於企業的業績成果有一定的影響力，因此，對於關鍵客戶的需求、對其專案的掌握狀況、客戶的滿意度，以及客戶未來發展的相關資訊，都是檢討的重點所在。

➡ Territory Review／區域檢討

各個區域與市場各有不同的特性、需求、發展狀況，以及需要原廠支援的重點方向，既然各有不同，原廠所擬訂的行銷策略就應因區域而有分別，因此，分別就各個區域的狀況來進行檢討，就是必定要做的年度檢討重點了。例如，各個區域市場所需要的產品類別會有所差異，因此，亦可針對各區域的「Top 10 Best Selling Products」／前十大熱銷產品來檢討、分析，由此即可訂出行銷的重點方向。

Forecast [ˋfor,kæst]

Forecast的英英解釋為「a statement about what is likely to happen, based on available information and usually relating to the weather, business or the economy」，其中提到了「available information」／已知資訊，而前面所說的Sales Review結果，就是available information，就是可用來做Sales Forecast的重要情資！從檢討所得來的資訊，加上企業所訂定的發展方向與目標，就可做出下一個年度的預估報告與執行計畫了。

Forecast的頭條一定會是業績目標預估，因為光是這個預估數字，可就得綜覽一堆業績數字，研究市場趨勢，訂出發展性與可行性兼備的業績總成長目標，再各別訂定出各區域、經銷商的業績目標…牽連了這麼複雜的工作，表示Forecast確實茲事體大，一定得內外考量、上下左右全面方析呢！

除了訂定業績目標之外，Forecast報告的內容還會包括擬訂的行銷策略、促銷專案、行銷工具，以及參展的計畫。蒐集充分的資訊，多方考量，才能制定最有可能成功的Forecast。正如有個「5P」是一路這麼P的：「Prior Planning Prevents Poor Performance」，有事前規劃就可以避免後頭的表現太糟糕，而Forecast就是帶領企業有步驟、有方法地朝目的地前進的指標，這是年度大事，辦成了這一件大事，就有可能成就更大的事業呢！

術語直達車，專業補給站

我們在最後這一節，要來好好說說關乎企業業務發展、管理與操作的「Sales Forecast」，Forecast要做個大概，還是要細細規劃呢？若這是個考題，想都不用想，當然要選「細細規劃」囉？只「做個大概」，雖然在實務上可能真的大概就是這麼做，但若有人問，怎麼好意思這麼說呢？既然知道這樣的粗糙會讓我們自己講來都不好意思，那我們就乾脆一點，不要粗糙，絕對要將Forecast做得仔細！

「The more detailed a forecast, the more accurate and practical it will be.」Forecast做得細，才愈有可能精確、愈實際可行、愈有可能達成。那精確不精確，有那麼重要嗎？有的！有兩件事跟你Forecast做得精不精確很有關係，那就是「manage cash flow」和「allocate resources」，有精確的Forecast，才能更精準地分配企業資金、人力與物力的投入流向，也能將火力集中攻向對的市場場域，如此一來就能大大提升成功攻下灘頭／拿下訂單的機率了。若是在一片混沌中擠出了一個漂亮的Forecast，但是一點也不實際，那麼再怎麼火力全開，也都難有效率、難以做出漂亮的投資報酬率來到達成功的彼岸啊！

Forecast要做到又細又精確，一定得好好打通歷史的任督二脈！啥歷史呢？就是歷史銷售資料呢！以前年度的業績分析就是Forecast不可缺的養分，另外還可以看之前年度的什麼資料呢？在此舉例說明如下：

➡ 成交案件客戶數 vs 失敗案件客戶數／The numbers of customers gained and lost

- 季節性模式 — 業績增減的週期／Seasonal patterns – period of sales rise and dip
- 各個客戶的訂單總額／The sales of each customer
- 特定產品的銷售狀態／The sales of specific product
- 各區域經濟情勢／District economic circumstances
- 採購的循環週期／Cyclical purchasing trends
- 突發狀況的影響／Impact of one-off events
- 行銷活動的影響／Impact of marketing campaigns

從這些方面來分析，對於找出客戶的採購習慣與模式有很大的助益，也有助於分析先前大訂單或大型專案是否在下一個年度還有重複採購的可能…掌握了銷售數據與資料的這些**market knowledge**，就更能有憑有據有理地預估市場的需求規模，繼而理出品牌本身在各個市場與各個區域的佔有率，然後，再訂出新的年度業績目標與成長率、制定行銷策略、設計行銷活動，以期達到目標。循此步驟所訂出的**Forecast**，得到的就會是有所本、有根據的目標，也會是合理的、可實現的目標呢！

尾聲、心聲

Last but not Least

好啦！現在到了這本書的尾聲了！看到這裡，你學到了哪些東西呢？「我學到了幽默風趣、正面思考、要體貼別人，還有評斷別人和評價自己時要有相同的標準！」……啊～不好意思，不小心把我自己的個性和一直以來奉行的圭臬寫出來了……哈哈！但這些態度在處理國貿業務與對外聯繫上有沒有相關？重不重要呢？如果你有認真看了我前面文章中的殷殷叮嚀，你就該知道，有人這樣問，當然要回答「重要！」哈～ 這種態度就對了！就像老闆抓了你問事情，你若是不清楚，能不能回答「不知道」呢？當然不行！請千萬把你要這樣回答的衝動壓下去，等到回座位上後再抓個東西吃，釋放因壓抑所產生的壓力！那要怎麼回答呢？請你問問自己，若你是主管，你問了下屬一件事，你希望得到什麼樣的回答？你一定也不想要聽到「不知道」嘛！你不想，別人當然也不想啊！那……要是真不知道，那要回答什麼呢？這時請你盡量說明一下你知道的任何相關訊息，若真沒啥訊息可回饋，就得要告訴對方你會怎麼去查、從哪兒詢問，找出答案後再來回報。這樣的態度才是處理事情的王道，若只回答個「不知道」，就一點兒也無法處理問題，絲毫也沒有表現出有要盡責、想要求知的態度。

說了這些，跟外貿業務有關嗎？有的！這樣的回應方式與處事的態度，對到任何事情皆合用，而對到業務與跨國溝通，一樣好用！我們在這本書說了這麼多，傳達的重點有二，一是英文的修煉，一是外貿知識的補強與深化。若是你真心想要讓你在外貿業務的溝通處理上更能駕馭自如，你會知道唯有多多吸納相關知識，才能讓你處事有辦法預應，遇事有辦法及時反應，有朝對的方向解決問題的本事。你若是已接觸外貿這圈子好些時日了，早已在外貿知識圈的領域中打滾了好一陣子了，那你怎麼能自絕於知識之外呢？「知識就是力量」！這句箴言一直就是「真」言，從1597年英國哲學家Francis Bacon這麼點明了這則真理之後，一直到現今的網路世代，這個真理一樣也是通往成功之路的關鍵！

除了看這本書補強外貿知識背景之外，還有什麼方法呢？喔！順道宣傳一下，有另一本書：《國貿英語溝通術》，裡頭有各個不同外貿操作程序中與國外聯繫的Email寫作情境範例、電話實例、單字與句型整理說明，還有國貿知識與國貿溝通的處理心法，可以讓你將國貿英文的基礎架設好，也可讓你對國貿知識有全面的認識。「為什麼獨獨介紹這一本書呢？」喔！問到重點了，因為這也是敝人在下我出版的書哩！歡迎一讀！

那除了讀這些書之外，還可從哪個方面來修煉自己的外貿專業呢？答案就在你手邊！你不是每天都會經手好幾種不同的業務相關資料嗎？說實話，你都有好好看過了嗎？看了，瞭解了，跟將它們拿來當學習的素材，在吸收知識的動力與成果層級上是有差別的！若是你碰到的是國外來的Email，它裡頭說到的業務操作相關知識，你都懂了嗎？若看到有的地方怪怪的，你有沒有辦法說得出癥結點在哪兒呢？若你不懂、不太確定，就請馬上查出、問出正確的方法。

美國教育學家John Dewey／杜威所提出的「Learning by doing」／「做中學」，之所以會讓人們老是拿出來中文英文都說一說，不是沒有道理的！關鍵就在於它的英文很簡單……啊不是啦！關鍵是在於這麼簡單的真理真的就是真正的道理！……哈！我想你知道我的意思……你工作上所經手的各式各樣的外貿相關文件與資訊，就是與你最有關的知識領域！我們要時時有好奇的學習心，不懂就查，也要培養自己的critical thinking／批判性思考或是分辨性思考的能力，訓練自己能分辨出、理出事物背後的邏輯，有疑處就查證，以強化、深化自己的外貿知識。

這本書說的是外貿英文、商用英文，所以要告訴你的是「英文很重要，非常重要，一定要通透到不行」嗎？我若是在這兒說「不」，那就是拿石頭砸自己的腳哩！英文是溝通的工具，既然是工具，那就請要求自己將它練成你的「利器」，這工具不能一直是鈍鈍的，不能幾年下來要割要切都還是使不上力，因為若是這樣，那就算你有滿腹的外貿知識與經驗，說不出口的就通通只能悶著往肚裡吞了，這樣不益健康，更有害你的自信，所以，英文這部分請紮實地從根基打起，一步步往上爬，總有一天連你做夢也會夢到英文版的夢境呢！

所以，「英文就是最重要的嗎？」我剛才有交待過了英文的重要性，所以現在我可以坦白的告訴你：「不」！啥？請放心，我會自己收尾的！英文重要，外貿知識重要，但是「態度」更重要！又有一句老掉牙的箴言要來囉：「態度決定一切」！雖然老掉牙，但它在我們每天的工作與生活上有太多太多的印證，許多思考的角度與行事的結果真的都關乎你處事的態度呢！要說外貿，要談業務溝通，若你有虛心向學的態度，能夠在面對問題時不光只是咕噥與抱怨，反而能遇事就有立馬分析、動手解決的態度，而且有站在對方立場思考的同理心態度，也有真誠待人處事的態度……若你有這些態度，哪裡還會有什麼事情是你處理不來、做不好的呢？！

做事就是做人，
期待我們都能盡心盡責做事，以誠以心待人，
都能在工作中讓自己修煉出更好的果實，成為更好的人！
共勉之！

國家圖書館出版品預行編目(CIP)資料

外貿業務英文-劉美慧著. -- 修訂三版. -- 新北市 :
倍斯特出版事業有限公司, 2023.07 面； 公分. --
(職場英語系列 ; 010)
ISBN 978-626-96563-5-6(平裝)
1.商業書信 2.商業英文 3.商業應用文4.電子郵件

493.6 112010499

職場英語系列 010

外貿業務英文-修訂三版(附QR code音檔)

初版三刷 2023年7月
定 價 新台幣380元

作 者 劉美慧
出 版 倍斯特出版事業有限公司
發 行 人 周瑞德
電 話 886-2-8245-6905
傳 真 886-2-2245-6398
地 址 23558 新北市中和區立業路83巷7號4樓
E-mail best.books.service@gmail.com
官 網 www.bestbookstw.com
總 編 輯 齊心瑀
特約編輯 James Chiao
封面構成 高鍾琪
內頁構成 菩薩蠻數位文化有限公司
印 製 大亞彩色印刷製版股份有限公司

港澳地區總經銷 泛華發行代理有限公司
地 址 香港新界將軍澳工業邨駿昌街7號2樓
電 話 852-2798-2323
傳 真 852-3181-3973

Simply Learning, Simply Best!

Simply Learning, Simply Best!